"*The Other Madisons* marks the culmination of Kearse's thirty-year investigation into not only her own family history, but also that of other enslaved and free African Americans whose voices have been silenced over the centuries." — *Smithsonian*

"With *The Other Madisons,* Kearse adds unvarnished truth to her family's official history, as well as America's . . . For a book that discusses the atrocities of owning people and the intergenerational trauma it causes, *The Other Madisons* is surprisingly charming and easy to read." — *Pasatiempo*

"Kearse's enlightening book, *The Other Madisons,* has not only been a labor of love for the author for thirty years but, more deeply, her life's purpose . . . Kearse's experiences with racism and those of her ancestors are deftly and sympathetically braided throughout the pages." — *Albuquerque Journal*

"*The Other Madisons,* as a thorough history of one family, may offer answers for other descendants of enslaved people as well. It is part personal quest, as Kearse works to understand and reconcile her own origins, and a carefully researched and documented correction to the American historical record." — *Shelf Awareness*

"Bettye Kearse's searing eye for truth educated, awakened, and stunned me. The heroics and pain of the author's kindred — descendants of slaves and a president — illustrates a family and country built on the shoulders of slavery. An unbroken line of ancestral oral history combined with Kearse's research illuminates ten generations, from slavery to the present, in a continuing battle against racism in which all Americans should fight. Kearse's generosity in presenting her hard-won truth is a gift I'll always remember with gratitude. I loved this page-turning book."

— Randy Susan Meyers, author of *Waisted* and *The Widow of Wall Street*

"*The Other Madisons* is a tale that Bettye Kearse was literally born to tell. Family lore held that she was the descendant of James Madison and his slave Coreen. How could she verify a history that existed outside of the historical record? As she journeys in search of her deepest, most painful family roots, Kearse unfurls an intensely personal tale that is also a quintessentially American story. Confronting colonialism and cruelty, power and its abuse, the silencing of slaves and the fraught complexity of intertwined nations and individual lives, *The Other Madison*s crafts a new kind of record, one that illuminates the power of a woman taking charge of her own truth."   — Paula Lee, PhD, historian and novelist

"Inheriting the role of *griotte* — family storyteller — from her mother, Bettye Kearse set out to preserve and deepen the knowledge about her family that oral tradition traces back to President Madison and an African slave named Mandy. As she travels to Virginia, Portugal, and Ghana, she shares with readers her research, her reflections, and her poignant emotional responses to her family's past. Her quest, at once personal and historical, is both engrossing and very moving."

— Gail Pool, author of *Lost Among the Baining: Adventure, Marriage, and Other Fieldwork*

THE OTHER MADISONS

The
OTHER MADISONS

The *LOST HISTORY of a*
PRESIDENT'S BLACK FAMILY

Bettye Kearse

Mariner Books
Houghton Mifflin Harcourt
BOSTON   NEW YORK

Library of Congress Cataloging-in-Publication Data
Names: Kearse, Bettye, author.
Title: The Other Madisons : the lost history of a president's Black family /
Bettye Kearse.
Other titles: Lost history of a president's Black family
Description: Boston : Houghton Mifflin Harcourt, [2020] |
Includes bibliographical references.
Identifiers: LCCN 2019024941 (print) | LCCN 2019024942 (ebook) |
ISBN 9781328604392 (hardback) | ISBN 9781328603531 (ebook)
ISBN 9780358505006 (trade paper)
Subjects: LCSH: Madison, James, 1751–1836—Family. |
Madison, James, 1751–1836—relations with African Americans. |
Madison family. | Mandy, active 18th century. | Coreen, active 18th century. |
African American families—History. | Racially mixed people—United States. |
Slaves—Virginia—History. | Freedmen—Texas—History.
Classification: LCC E342.1 .K43 2020 (print) | LCC E342.1 (ebook) |
DDC 973.5/10922—DC23
LC record available at https://lccn.loc.gov/2019024941
LC ebook record available at https://lccn.loc.gov/2019024942

Book design by Chloe Foster
Family tree by Carly Miller

Printed in the United States of America
DOC 10 9 8 7 6 5 4 3 2 1

The chapter entitled "Destination Jim Crow" was first published under the same
title, in different form, in the Fall 2013 issue of *River Teeth Journal*.
Reprinted by permission of the author. All rights reserved.

Image credits appear on page 249.

To my mother, Ruby Laura Madison Wilson,

who taught me to value pride

To my father, Clay Morgan Wilson III,

who taught me to value humility

To my grandfather John Chester Madison,

who taught me to value a story well told

Madison Family Tree

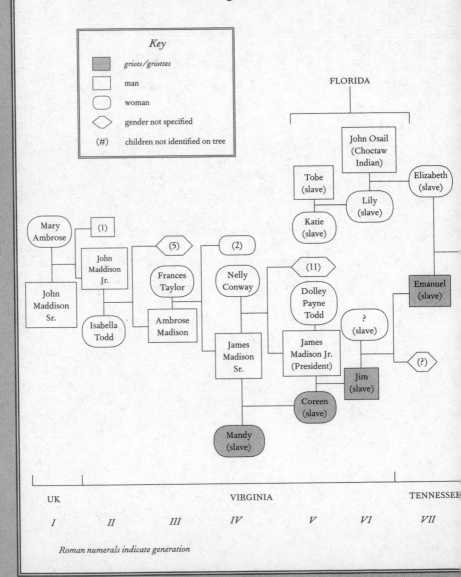

Key

- ▦ griots/griottes
- ▢ man
- ◯ woman
- ◇ gender not specified
- (#) children not identified on tree

FLORIDA

John Osail (Choctaw Indian)

Elizabeth (slave)

Tobe (slave)

Lily (slave)

Mary Ambrose

(1)

John Maddison Jr.

(5)

(2)

(11)

Katie (slave)

Frances Taylor

Nelly Conway

Dolley Payne Todd

? (slave)

Emanuel (slave)

John Maddison Sr.

Isabella Todd

Ambrose Madison

James Madison Sr.

James Madison Jr. (President)

Jim (slave)

(?)

Coreen (slave)

Mandy (slave)

UK

VIRGINIA

TENNESSEE

I II III IV V VI VII

Roman numerals indicate generation

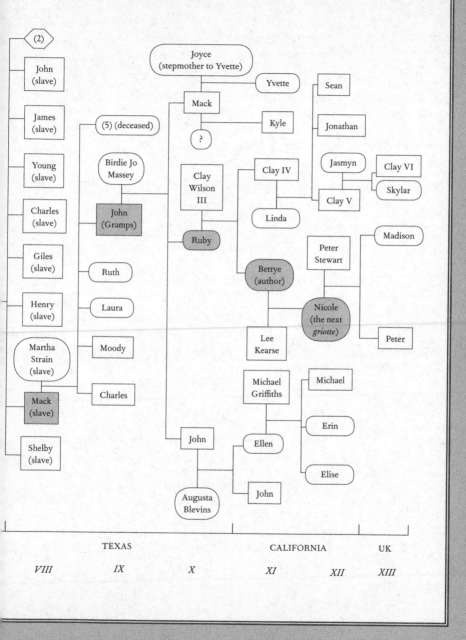

TEXAS   CALIFORNIA   UK

VIII   *IX*   *X*   *XI*   *XII*   *XIII*

Contents

THE OTHER MADISONS

Prologue

I am *griot,* master of eloquence, the vessel of speech, the memory of mankind. I speak no untruths. This is the word of my father and my father's father. Listen to me, those who want to know. From my mouth you will hear the history of your ancestors.

— West African *griot* opening chant

For thousands of years, West African *griots* (men) and *griottes* (women) have served as human links between past and present, speaking the ever-expanding stories of their ancestors and the history of their people — accounts of births and deaths, conquests and defeats, times of plenty and times of famine, vast empires and small villages, nobles and heroes and commoners. These men and women are not simply oral historians, genealogists, storytellers, and teachers; they are also spokespeople, exhorters, interpreters, judges, poets, musicians, and praise-singers. Their wisdom and their voices preserve not just a family or a community but an entire culture and its values.

In his book *Griots and Griottes: Masters of Words and Music,* Thomas A. Hale writes, "No other profession in any other part of the world is charged with such wide-ranging and intimate involvement in the lives of people." These "wordsmiths," he continues, are the "social glue" of society. Their words and the layers of meaning behind those words influence how each person views himself in the present and on the continuum of past and future. In the eyes of West Africans, *griots* and *griottes* are fundamentally different from other human beings. Even their burial rituals are unique. Though neither religious icons nor sorcerers, they hold an aura of power and mystery that makes them at once frightening and revered.

American slave owners successfully abolished many African customs, but the tradition of oral history has held strong. For many African-American families, including mine, this tradition is all that preserves the legacies our ancestors left for us. In the "official" history of America, their stories were excluded, ignored, marginalized, or distorted. But in each generation of my family, the *griot* has kept the stories alive and added his or her own important lessons and personal tales to the saga in order to leave evidence that they, like their predecessors, existed and, though often confronted by restrictive circumstances, did all they could to make the most of their lives.

Our first wordsmith was a slave called Mandy. When it came time for me to take on the role, she and our family's other *griots,* living and dead, helped me to discover and add my own lessons and personal tales. Their words encouraged me to write down their legacies and include my own for the coming generations. But it was Mandy who held me up when I doubted I could become the *griotte*. Sometimes I felt so close to her I could hear her voice. It was like a xylophone: precise, clear, musical. The melody's lilt slid down at the end of each sentence, the consonants percussive, the vowels soft — the inflections of the Ga language of Ghana.

Mandy

When I was a girl, I didn't know people stole people. I used to sneak away from my village and go to the edge of the ocean, my ocean. I was very young and a little foolish then. I thought that huge body of water belonged to me. A big, knotty tree with twisted branches stood alone on a hill, where it watched over my village on one side and the water on the other. My favorite spot was a cove hidden among tall boulders. I went there whenever I could. All kinds of reptiles, insects, and sea plants clung to rocks, slipped into cracks, or hid in shadows to get away from the sun and wind pounding the beach. Sometimes, I took the small creatures home, but usually, I left them where they were so they'd be there whenever I came back.

Even if I was supposed to be tending chickens, cooking, or watching my brother, I'd sneak down to the water. Sometimes I could only stay a minute, but sometimes I stayed for hours, digging my toes deep into the cool sand. Warm sea wind brushed my cheeks while twinkling blue water hurried to the shore and curled into white, foamy ringlets that pulled the sand toward the bottom of the ocean. When the sand drew away with the water, I dug my toes in further, because

I could feel it tugging my feet, trying to take me with it too. But I thought I was going to stay on that land forever. I grabbed the sand with my toes like they were the roots of a tree, and I stayed right there. Funny how my little toes were stronger than that whole big ocean. I didn't move but a tiny bit.

In the mornings, the sun was hot on my shoulders. In the afternoons, it was my forehead and chest that tingled in the heat. Sun shining down on my body, I watched the water get higher or lower, so slow I couldn't see it changing. Bubbly water hid thin strips of sand, but if I kept watching, shellfish I hadn't seen moments before scurried across the shore, trying to keep up with the ocean and leaving behind smooth and shiny pink or silver or green or blue or rainbow-speckled pebbles.

The biggest part of the ocean, the part that touched the far-away sky, had too-many-to-count white peaks that grew smaller and smaller in the distance. In the coves, the water calmed into soft ripples, like ribs rising and falling in regular breaths. On the open beach, frisky waves ran up to the bank, hit the rocks, then splashed into white sprays, just like they were playing. Any movement or sound in one part of the ocean swelled up or hushed down in another.

Sometimes when I got too warm, I'd slip into the cool water — real slow, to give it a chance to know I was there. Then I'd let the ocean pull me in, lift me up, push me down, like I was part of it, a most powerful and peaceful feeling. Afterward, I'd sneak back home, salty and wet, my hair sparkly with sand.

So much has happened since then, but I will never forget the ocean near my village.

One day, I was sitting on the tallest rock near the cove when I saw the water looking like it was gasping for breath, pulling itself down, hard. I thought a storm was coming, but the sky was clear and calm and blue. Far away, I saw a boat, getting bigger as it came, churning up waves.

I hid between some rocks. I stayed there for a long time, listening to harsh shouting, sorrow-filled wailing, boots thumping, fright-choked gagging, and metal scraping against stone. Finally, when the sky was dark and the horrible noises had stopped, I crawled to the top of the hill to hide in the twisted tree. It glowed in the moon-light, quiet and still, like nothing had changed. But just as my fingers touched the trunk's solid bark, someone grabbed me and threw me to the ground. My head hit a shallow root.

Hands, lots of hands, grabbed my arms and my legs and my neck. Then one pair of hands, black hands, crushed my chest so hard I could hardly breathe. The tall man whose hands those were hissed ugly things in my ear and dragged me over tangles of gnarled-up roots, down the hill, and across rocks and sand. Then he picked me up and dropped me into a small boat filled with tied-up people. I got bound up too.

The boat rocked and bumped along the coast, and when the sun came up, I saw a huge white building on the edge of the water. The boat stopped, and the man pulled me out.

All I had with me were the pretty red beads in my hair, so I thought he would take me back to the tree. He didn't. When we got inside the big building, he pushed me down onto a stone floor covered with damp, foul-smelling dirt. A horde of crying, screaming, shouting people pressed in on me when they tried to move. One small window, way up high, cut a sliver of dust-filled light. The thick air smelled like rot itself. I didn't try to talk to anyone. I didn't see anyone from my village. Everyone was a stranger. All I could do was cry.

Many days later, all of us, children and a lot of women and men too, had to walk or crawl through a gate, across the sand, and up onto a boat much bigger than the one that had carried me already so far from home. Someone shoved me down a ladder and onto a plank of splintery wood. I tried to sit up but banged my head against the floorboard above me.

The boat began to sway. Splinters dug into my legs and hands. Rats crawled over my feet. I forced my tears away and called out for help. I called again. The only answers were pleas like mine. Nearby, someone was sobbing. When I reached toward the sound, I dragged someone else's hand with mine. My wrist was chained to another girl's wrist. We tried to speak to each other. I couldn't understand her; she couldn't understand me. I felt completely alone.

I didn't know the word slave back then, but I knew I had no chance to be free again. No home. No mother and father, no big sisters and baby brother, no dances and drums, no lessons from the village elders, no friends to laugh with, no grinding grain with the women and other girls, no chickens clucking and running at my feet, no big, twisted tree. No cool, mighty ocean.

I knew what the white men on the boat were doing to the women, the boys, and girls like me, so I kept watch, every moment, for my turn. But I was lucky — then.

It happened in the place where they took me to live, to work. Late one night, Massa found me alone in a cabin. He didn't stay long but in that short time, he tainted the woman I was meant to be. After he left, I bore so much anger I couldn't believe it was me feeling that way — angry to my soul.

Somehow, as time went on, I reached deep inside myself and pulled up the lessons my mother and father and the elders had taught me. I knew that, no matter what, I could bring honor to my family and our ancestors. *I figured out how to use my anger. I still grieved for my old life, sometimes wished I was dead. But I couldn't let anybody destroy me. Not anybody. I made a new life for myself.*

　But when young Massa hurt my baby Coreen — oh God!

　I cried when she looked at her image on the surface of a pond and then reached in with violent slaps to fracture her reflection. When she started throwing rocks at the defiled woman she saw, I fell to my knees. I knew what she was feeling. I knew Coreen was exploding with helplessness and hate. From the day she was born, there was never a thing I could do to protect her, but after that man damaged her, I took her into my arms and rocked her and talked to her, just as I did when she was the sad little girl who hung on to my legs. "Anger can tear you up," I said this time, "but never forget — never — if you're fighting mad, feeling that way can give you strength and keep you going."

I

The New Griotte

President Madison did not have children with his wife, Dolley. Leading scholars believe he was impotent, infertile, or both. But the stories I have heard since my childhood say that James Madison, a Founding Father of our nation, was also a founding father of my African-American family.

According to the history told by eight generations of my family's *griots,* Madison had a relationship with one of his slaves, Coreen, that resulted in the birth of a son, Jim, who was sold and sent away when he was a teenager. Jim was my great-great-great-grandfather.

My earliest recollection of hearing this story was as a five-year-old attempting to stand still while my mother worked on the dress she was creating for me. Every time I had a piano recital, she sewed me a new dress, and every time she sewed for me, I became bored and fidgety. I dreaded the fittings more than the performance itself. The performance lasted little more than two minutes; the fittings took *forever.* Mom pinned a seam; I tried on the dress. Mom sewed a seam; I tried on the dress. Mom pinned a hem; I tried on the dress . . . It was torture. My mother designed my outfits, but I did not care about ruffles, lace, and satin trim, and I did not want to play the piano. I wanted to dance. The closer the dress came to its final shape, the closer I came to driving my mother to her wits' end. At every opportunity, I'd slip away and dance to the music in my head. I loved the *Nutcracker Suite*. I was the Sugar Plum Fairy. I arabesqued, twirled, pliéed, then twirled again, careful not to let the pins stick me. But my reprieves were brief.

"Dolly, come back," Mom would call. "I have the box out. I'll tell you your favorite family stories." And if I kept whining and squirming, Mom would throw up her hands and say, "Please, Bettye, why do I have to keep reminding you? Always remember — you're a Madison. You come from African slaves and a president."

These words have never been for me alone. They have guided my family for nine generations and evolved to meet the demands of America's changing times. In the antebellum years, my enslaved ancestors used Madison's name as a tool to help them find family members who had been sold and sent away. During Reconstruction, the saying inspired my ancestors to make the most of their lives, now that they were free. And since the Jim Crow era, it has reminded us that our enslaved ancestors were strong, remarkable people.

When I was a child, I thought the directive was merely Mom's way of telling me to behave myself. In part, it was. She employed the exhortation to set the standard for my conduct in many of my childhood moments, good and bad. As I grew into adulthood, she taught me to incorporate it as the standard for how I should live my life. My actions should reflect both my presidential ancestry and my pride in knowing that the blood of slaves runs through my veins.

==

But it took time for me to learn how to live up to my legacy and figure out what it meant to me. I am seven years older than my brother, Clay Morgan Wilson IV, nicknamed Biff. Because of the difference in our ages, our parents often joked that they were doing the best they could to survive the challenge of raising not one only child, but two. Our parents, Dr. Clay Morgan Wilson III and Ruby Laura Madison Wilson, enrolled us in good public schools, paid for music lessons — and dance classes for me — took us to cultural events and church, sent us to summer camps, and taught us how we should behave. Their only demand was that we do our best at all times. We didn't have to *be* the best; we only had to *do* our best. My mother would often take me aside and say, "All you have to do is to be sweet and smart." I didn't understand what being sweet had to do with being smart, and I didn't understand why my brother was told only to be smart. I was her sweet doll for whom she made fancy, one-of-a-kind dresses, and she called me "Dolly."

When I was seventeen, my mother, backed by my father, insisted I be a debutante in the cotillion. I acquiesced; otherwise I would not be allowed to go out on one-on-one dates until I started college. It was a choice that was not a choice for a seventeen-year-old, and I knew that many of the other potential debs were already dating and had serious boyfriends.

The cotillion is an annual ball sponsored by the Bay Area chapter of the Links, which, when I was a deb, was a national organization of wives of prominent black men (today it is the women's accomplishments that qualify them for membership). Dating back to the seventeenth century, the debutante ball was originally a European tradition that declared young women with the "right" credentials eligible for marriage. The goal was for the girls to snag husbands.

In 1960, I and most of the other debs were thinking about college. Few of us were interested in marriage. But on the night of the cotillion, one by one, each deb, wearing a long white dress evocative of a wedding gown, stepped onto an elaborately decorated platform to be shown to society.

Me at age seventeen, Links Cotillion, 1960

I stood in the wings, more resigned than nervous, waiting to hear my name called. When my turn came, I lifted the hem of my dress and climbed the steps up to the platform. There I stood, framed by a plethora of white flowers and ribbons, the spotlights blinding me, as the announcer read off my lineage and described my accomplishments. I smiled hard and curtsied deep. I couldn't see the audience, but I heard the applause validating my worth.

Unaccustomed to high-heeled shoes, I gingerly stepped down a red-carpeted stairway. My proud, tuxedoed father met me at the bottom and paraded me around the ballroom. Then my proud-of-himself, tuxedoed young escort paraded me around the ballroom. When the last of the debs had gone through this ritual, we formed three circles on the dance floor to perform a rigidly choreographed minuet. We were petals on a flower. Black girls in white.

We flaunted our American middle-classness, I now realize, and gave no thought to the time, a hundred years earlier, when our ancestors stood on platforms to be appraised and parceled out.

Later, a photograph of me receiving a certificate for best academic achievement appeared in *Jet* magazine. I had done a good job of representing my father's family, the Wilsons, and my mother's family, the Madisons. But I thought the whole thing foolish. There was pressure not just to achieve but to broadcast that achievement. I resented being put on display.

Oakland had a racially and culturally diverse population, and I could take a train or a ferry across the bay to San Francisco to explore the museums and shops, attend plays and dance performances, or meander through the long, narrow park that ended at a zoo with the Pacific Ocean a short walk beyond. San Francisco was one of the most sophisticated, pretty, and glamorous cities in the world, but it was too close to home.

The family saying anchored me — I knew who my ancestors were and what they had accomplished; they had set examples for me to follow. But it also felt like a trap. It sounds banal, but I wanted to explore life unencumbered by an overprotective family and a watchful black middle-class community. I knew that whatever I wanted to try, I could find it in New York City. I also knew my parents would let me go anywhere to further my education. New York University was the perfect choice.

I attended class faithfully and fulfilled the academic requirements, but the city around me beckoned. I arranged my schedule so I could attend Wednesday matinees on Broadway. I saw Rudolf Nureyev and Margot Fonteyn dance as Romeo and Juliet; Sammy Davis Jr. act, sing, and box in *Golden Boy;* Sidney Poitier become Walter Lee Younger in *A Raisin in the Sun*; and Barbra Streisand and Sydney Chaplin portray Fanny Brice and Nicky Arnstein in *Funny Girl.* Judith Jamison was the dancer I had dreamed of becoming almost two decades earlier. Her performance in *The Prodigal Prince* with the Alvin Ailey dance company was so mesmerizing it made me cry. I loved the subway; for a nickel, it could take me almost everywhere in the big, noisy, exciting city.

One afternoon, I attended a talk by science fiction writer Ray Bradbury, who urged the audience of NYU students to "muddle around in life." Only by exploring life's ups and downs, he explained, could we grow as young adults. My family's expectations were clear and binding; there was little room for me to muddle, but I gave it a try. I partied with friends in disco after disco until closing time. We often ended up in someone's apartment, dancing until early morning. I remember sitting on the floor at a friend's smoke-filled place as he tried to teach me what to do with a joint. I failed. It made me feel weak and dizzy and triggered my asthma, but I was delighted I had tried. My parents would not have approved of my efforts, tame as they were, had they known what I was up to.

Well before graduate school, I'd found that being a descendant of African slaves had a significant impact on my life. Being a descendant of a president, however, did not. I came to resent the reverence for James Madison the directive demanded. He was a Founding Father and a president, but he had also owned people. And I did not believe he should be excused merely because, at that time, having slaves was the norm among landowners and the wealthy.

It would take me years to articulate this to myself and then decades to explain to my mother that the saying she revered and lived by echoed with the abuses of slavery.

In 1990, at the age of forty-seven, I had my own pediatric practice and had been married for twenty years to an accomplished physician-researcher. Our daughter, Nicole Elise, was a bright, beautiful, often hilarious seventeen-going-on-seventy-year-old who described herself as "spoiled but not *rotted*." She loved to burst into the house shouting, "Hi, honey, I'm *black!*"

Life was good. But one day, my mother called, sounding tired. "Dolly, I'll be coming east soon to bring you the box." Mom's pet name for me was a reference to that little girl she once dressed up in organdy and lace to perform in piano recitals and to the elegantly gowned teenager who smiled and curtsied in the debutante ball. Mom was not referring to our ancestor's wife. She had no fondness for Dolley Madison.

I knew which box she meant, and I was stunned that she was giving it to me. The taped-up, saggy cardboard box, big enough to hold four or five of my childhood dolls, contained priceless family memorabilia — photographs, land deeds, bundles of letters, wills, and birth certificates. A smudged copy of an 1860 slave census listed my great-great-grandparents and their ten children by gender and approximate age.

When I was a child, Mom showed me the list and told me, "That male slave there, age fifty-two, and that female slave, age thirty-seven, were your great-great-grandparents, and that male slave, age nineteen, was your great-grandfather. The others were his brothers." When I asked whether the slaves had names, my mother answered, "Of course they did, but most plantation owners thought of slaves as belongings, like tables or jewelry or mules. Any favorite possession might get a special name, like the mule Ol' Maizy. Your great-great-grandfather's name was Emanuel, your great-great-grandmother was Elizabeth, and your great-grandfather was Mack."

I remember my youthful fascination with the photographs. A few were in color, most in fading sepia, some pasted on velvety black paper in albums, and others left scattered in the box. Many of the subjects, especially those in the older pictures, appeared threatening as they stood or sat stiffly in funny-looking clothes and glared right at me.

Mom had never mentioned that I would be the next *griotte*. There had been a *griot* in every generation in my family for two centuries; my mother was the seventh one. And now, it seemed, I was to be the eighth.

After my mother phoned to tell me she was bringing me the

box, I tried to read a novel but soon put it aside. I picked up my knitting but quit after two rows. I stood on the back porch trying to persuade myself to throw down some mulch in the garden or trim the hedges, anything other than dwelling on the possible meaning of Mom's upcoming visit. When my husband, Lee, asked what was wrong, I could only shake my head. After many anxious hours, I decided I would inquire about her health the moment I saw her. I had to know why *now* was the time for me to take the box into my care.

Compounding my concern about Mom's health was my terror about becoming a *griotte*. Understanding our directive, "Always remember — you're a Madison. You come from African slaves and a president," had presented a daunting challenge for me from the time I first heard it, and it was even more of one now that I would be responsible for ensuring that the torch of family pride and history would not go out. I had tried to live in accordance with the directive but worried I would never be the Madison my family expected me to be. I always felt embarrassed when my mother introduced me to someone. She would say, "This is my favorite daughter" — the family joke was that I was her *only* daughter — "Bettye. She's a double doctor. She graduated from UC Berkeley, then she went to NYU, met her husband, and got a PhD in biology. She worked in a lab for a while but didn't like doing the same thing over and over, which doesn't surprise me. Two years after her daughter, Nicole, was born, she decided to join her husband in med school at Case Western Reserve University in Cleveland. It's one of the best schools in the country. Now she's a pediatrician in Boston and has her own practice!" As this litany rolled on, I wanted to disappear, but I understood that, for my mother, my credentials were successes not only for me but also for the family legacy and for herself, the girl who had picked cotton in small-town Texas.

In school, I had learned a lot about President Madison, and my mother had told me stories about my enslaved ancestors, but to be the *griotte* my family deserved, I would have to enter my forebears' lives and do my best to understand who we were as the "Other Madisons." Our archives were precious, the stories inspirational, and the storytellers wise. I did not feel wise. For all of my degrees, I had only skimmed the emotional depth I'd need to face the truth head-on and tell my family's story. What would I add to the family *griot* tradition?

My grandfather John Chester Madison had contributed many lessons. He was the sixth of our family *griots*. I called him Gramps. His father chose him to be the *griot* of his generation because he cared deeply about history, especially family history, and because he was a gifted storyteller who could make our ancestors as real as if they were about to join us at the dinner table.

When I was six years old, Gramps flew from Texas to California to spend Christmas with us. One evening, he sat me on his lap and told me his version of the traditional Negro folktale "The People Could Fly." His face was smooth and golden brown, his mustache and wavy hair white. Tiny moles stippled his cheeks and neck like freckles. Round, silver-rimmed glasses sat precariously on the end of his nose but never slipped off, not even during the most animated parts of his storytelling. I snuggled in to listen.

"The people were black and strong and magic," Gramps began. "More than magic, they were wondrous. They were God's people. Some folk way over in Africa could step up into the air and fly away by their wings. Yes, our people had wings back then. Sometimes a whole bunch of them flew together, looking like a plume of butterflies rising here, then there in the big, blue sky. Every day, they would fly someplace to fish, fly back home, feed their families, fly away again, find some berries, fly back home with the berries, and raise their babies. That's the way it was for many, many years.

"But one day, some white men with whips and chains came from across the ocean and shoved the people into ships and stole them far away. The black folks were so aggrieved that their wings fell off and turned to dust. The dust flew into their eyes and made tears.

"Out in the hot, burning fields of the new land so foreign to them, the folk from Africa worked from before sunrise to after sunset, from 'can't see to can't see.' The harder they worked, the more they forgot about being strong and magic, and misery made them forget about their wings.

"Years went by. Then one morning, an old black man came strutting down the road, swinging a hoe. He was tall and had a long white beard and long white hair that hung around his smiling, black face. His clothes were the same as what the people working the fields wore, and nobody thought there was anything special about him. Except his smile.

"He walked over to a young woman. She was working so hard her legs shook and working so fast, she couldn't stop to feed the baby crying at her feet. She was pretty, but her dark brown face was streaked gray with dry, salty tears, and her eyes were puffed red and sad.

"The old man said, 'Mary,' and the woman looked up from her work, surprised he knew her name.

"'Mary,' he repeated. 'It's time. Ain't safe fo' yo' li'l boy.'" Gramps said this in his slave voice, as if he were the old man in the tale. "'Ol' Massa go'n hurt him. Best you go right now.'

"Mary wasn't sure what he meant, so he explained. ' 'Member yo' magic. You gotta 'member now, befo' it's too late.'

"Then he whispered some secret words into her ear. She stared at him, and her eyes turned bright. She did remember her magic! Mary stuck her hoe into the ground, bent down, and scooped up the baby. She stretched her neck and began to float toward the sun. Right after her feet left the earth, big silver wings sprouted from her back, and she soared up high, smiling and hugging her baby to her chest. Mary looked down and saw her hoe standing and trembling in the ground, right where she had left it. The massa of the plantation couldn't believe his eyes.

"Next, the old man went to Joseph, who was chopping and hoeing so hard, it seemed like the flesh might fall off his bones. He didn't see Mary and the baby in the sky.

"'Joseph,' the old man said, 'it's time.' Then he whispered something to Joseph and pointed to Mary and the baby. Joseph smiled real big.

"When Massa saw this, he ran up and threw a long whip around Joseph's legs, cutting them, making them bleed onto the dirt. Joseph pulled that evil strap from around his legs, shoved his hoe into the ground, jumped into the sky, and flew over the fields. His silver wings cast a broad shadow, and as he flew, blood dripped from his legs and killed the tobacco.

"Now Massa was really mad, especially when he saw Joseph's and Mary's hoes standing in the middle of the empty, black field, shaking like they were laughing at him. Massa stomped his feet and hollered, but Joseph just flapped his wings and flew higher until he caught up with Mary and the baby.

"Soon, one by one, folks toiling in the fields heard the old man's magic words and took to the sky. With all of them heading toward the sun, the sky started to get dim. Below them, the fields withered, and the ground opened up in big crevices. Now there were so many hoes standing and quaking that the earth began to rumble, and the cracks got wider and deeper.

"Finally, the old man lifted himself up into the air and flew over Massa, who started quivering because the ground was moving, because the spread of the old man's wings shut out more light and heat from the sun, but mostly because he was mighty scared. Then the old man flapped his wings so fast the sky turned into a wild wind that pulled all the trees out of the ground and threw them off the plantation.

"Massa lay down on the dry, cracked earth so the wind wouldn't throw him off too. The old man looked down on Massa but didn't say a word. He just flew higher. When he got higher than the mountains, his ragged pants and shirt began to shimmer and glow, turning into flowing, white robes. His hoe turned into a staff of pure gold. He was a glorious angel sent by the Lord!

"As the angel flew toward the sun, the sky got blacker and blacker. Massa kept his face hidden in the ground, so he didn't see that when the angel got close to the sun, it shrank down and disappeared. Then, where the sun used to make its gold light in the east, a ball of silver light began to grow, silver just like the wings of God's special people. That light grew and grew, and filled the sky. It was a big, beautiful star, the Star of Bethlehem. But Massa didn't see it because he was still lying on the ground, shaking and hiding his face. Nobody paid no mind when a giant gap opened wide under Massa and swallowed him up forever."

After Gramps told me that story, he lifted my chin so I was looking into his eyes. He said: "In a way, baby girl, we all have wings. We just have to know what we want, figure out what we have to do to get it, and then do it. Christ showed us how. Nobody's wings are any stronger than ours. That's the way God made us. We're the Other Madisons."

Later, as a forty-seven-year-old in need of reassurance, I could see his eyes and hear his voice.

Two months after my mother told me she was giving me the box, she flew from Oakland to visit me in Boston. When she stepped off the plane, I saw nothing out of the ordinary about her, but I still searched for clues. Three decades earlier, my grandfather, aware he was dying, handed over this important role to his daughter. Was Mom ill? Did her visit mean I was about to lose her? Was this the way it worked?

Mom had on her red hat — a good sign, I decided — and she carried a carousel of slides and the saggy, taped-up cardboard box. The carousel was nine years old; the cardboard box, several decades older. My mother's visit would last three short days, but it would change my life.

The ride home from the airport was uneventful, and as we pulled into my driveway, the evening air, chilly in the aftermath of a late-autumn storm, was soothing. Around my house, trees with sparse leaves clinging to branches stood in black silhouette against the twilight sky. Their ethereal beauty reassured me.

Mom entered the living room and removed her hat, revealing the smooth outline of her head through a silver mist of frothy hair.

Her cautious gait and rounded shoulders made me aware that she had aged in the few months since I had last seen her.

She took a hand mirror from her cosmetics case and turned her back to the large one on the wall. Then, as I had seen her do hundreds of times before, Mom adjusted her head and angled her wrist to create a looking-glass corridor. She stood gazing silently into the hand mirror for several moments, watching her image become smaller and smaller along an infinite reflective portal.

What did she envision far down the passage? The young woman she used to be, the taut, golden skin framed by an arc of thick black hair? With the purpose of her visit in mind, did she imagine our ancestors and descendants lined up along the passage, slaves and their masters and the generations before them standing at the far end, and children, grandchildren, and future great-grandchildren at the other?

Mom lowered the hand mirror, leaned forward, and placed it between the slide carousel and the colorful hat on the coffee table. The box sat on the floor beneath the wall mirror.

"How are you?" I asked.

"Getting older, of course, but I'm okay ... I decided to bring the box rather than call you to come get it because I didn't want to worry you. I remember my father's call all too well."

"But why now?"

"I want to give you plenty of time to write the book."

2

The Box

Before there was a box, there was a Bible stuffed full of family memorabilia. It was ordinary to look at, black except for the rough, tan leather showing through dry cracks along the edges of the binding and on the ridges around the faded gold letters that read *Holy Bible*. Gramps's special feelings for the Holy Book had grown not just from devotion to Christian ideals, but from pride in the family documents he had painstakingly stored within its protective cover.

But he lost it. In 1940, he accepted an assignment to become principal of a new school, and Gramps moved, with his wife and three young children, from one small Texas town to another. Before leaving his old home, he wrapped the Bible in newspaper, tied it with string, and placed it on a bookshelf that was to be sent to his new home. The bookcase arrived at the new house. The Bible did not. Gramps was distraught. He assumed that his most cherished possession had been mistaken for trash, and he never forgave himself for not being more careful. He had lost part of who he was and felt he had failed as a *griot*. Gramps never forgot the family saying and continued to tell the stories. He purchased another Bible and planned to seek out keepsakes to save in it, but day-to-day responsibilities usurped his time, and time ran out too soon.

Seeing how devastating the loss was to her father, Ruby began gathering family mementos in a cardboard box. Gramps died in 1960. My mother, then forty-two, felt she should have done more. Gramps's stories, his messages, his love for his ancestors, and the legacy he had sought to preserve had to be kept alive.

Sixteen years after Gramps's death, the box was less than half full. That was 1976, the year Mom read Alex Haley's novel *Roots*, the inspiration for the historic television miniseries that aired in 1977. Haley's work helped many African Americans realize that their family histories were not only important but accessible. Mom decided to do something about recovering hers.

She left her home in Oakland, flew to Salt Lake City, and combed through genealogical archives at the Mormon Family History Library. The collection was vast, but Mom did not find what she was seeking — evidence that our enslaved ancestors had existed. From Utah, she traveled to Virginia and visited Montpelier, President James Madison's home. At that time, the staff was meager and no one had begun exploring the roles slaves had played on the former plantation, so she spent hours at the Orange County Courthouse, searching rolls of microfiche. She did not find the names she was looking for. Several weeks later, after a brief stop back in Oakland, she was off again, this time to Bastrop County, Texas, where the black Madisons lived at the time of emancipation.

When she was not on the road, Mom spent hours in libraries, worked the telephone at home, and wrote relatives throughout the country. After five years, she had amassed scores of old letters, photographs, and copies of most of our lost documents.

The only enslaved ancestors she had found records of were her great-grandparents Emanuel and Elizabeth and their eight sons, but my mother had become both *griotte* and archivist. She was satisfied that she had filled the cardboard box.

She created a slide presentation, and throughout the 1980s, she shared the stories and what she had found with family, friends, and historical and genealogical organizations in California, Texas, Massachusetts, New York, and Virginia. My brother and I teased her about "The Black Madisons' Lecture Circuit," but her commitment to keeping our family's story alive was evident every time she carefully donned one of her colorful hats, waved to us from the car, then backed out of the driveway on her way to each speaking engagement.

I saw her presentation for the first time in the darkened living room of my parents' home. With my ten-year-old daughter, Nicole, and a few other family members and close friends looking on, my mother stood next to the slide projector, advancing the slides one by one and telling the story behind each image as it appeared on the makeshift screen, a bed sheet taped to a wall.

The first slide showed a map of England. "The story of the Other Madisons began in the seventeenth century," Mom said as she pointed to a small dot on the map. "This is where John Maddison — the name was spelled with two *d*'s then — grew up. He was the president's great-great-grandfather." The next slide displayed a map of Africa. Placing the pointer on the coast of an area that is now Ghana, she said, "And this is about where slave catchers found Mandy. She was the matriarch of our family." A few slides later, a picture of a solemn, formally attired man appeared on the screen. "This," Mom said proudly, "is President James Madison. And this," she said, a slight edge to her voice as a slide of an elegantly dressed woman appeared, "is his wife, Dolley. If it had not been for Dolley, known for her fancy clothes, expensive tastes, and elaborate parties, Jim, Mandy's grandson, might not have been sold and lost to us."

Several images later, a tawny-skinned, white-haired man appeared on the screen. "This is my grandfather Mack Madison. So stately and handsome in his suit and vest — who would guess he had ever been a slave?"

Mack Madison (1837–1912)

Next was a picture of a fair-skinned woman wearing glasses and a lace-trimmed blouse. Mom said, "This is Mack's wife, Grand-mother Martha. Together they had ten children, but only five lived to adulthood."

Martha Murchison Strain Madison (1842–1914)

The following slide was a composite. On the right was a studio portrait of four young adults, two men standing behind two seated women, all of them carefully groomed and wearing starched and

Mack and Martha's surviving children. Left photo, Charlie;
right photo, standing from left to right, Moody and John Chester
(Gramps); seated, from left to right, Laura, and Ruth

pressed attire. On the left was an image of a man in a jaunty hat
and a crumpled suit.

Mom explained, "These are Mack and Martha's surviving children. The four together are Laura, Ruth, Moody, and my father,
John Chester. The man standing alone is their brother Charlie. A
few years before these pictures were taken, Charlie got into trouble with some white men and had to leave town. He even had to
change his name; John Miller is what he decided to call himself.
He was afraid of getting lynched . . .

"Here is another picture of my father," Mom said, displaying another slide. "Isn't he good-looking! He wanted to be a doctor, but one thing after another got in the way, so he became a teacher and then a principal."

John Chester Madison (Gramps) (1882–1960)

Many slides later, the show ended with a picture of Nicole when she was a coy two-year-old.

Mandy

By the time I was stolen, the pain of the cutting of my woman-to-be parts was only a memory. The ritual had prepared my body for the husband I would someday honor with children. I had become a proud young woman, tall, almost as tall as my father. Some girls didn't like the scars the elders had made on their cheeks, but I liked the way mine felt whenever I smiled. My mother's scars were similar, and the tightness in my face reminded me of how much I looked like her. I thought my mother was the prettiest woman in the village.

She knew how to make my hair beautiful. On the first morning after every full moon, I used to sit on the ground between my mother's knees while she braided my hair. As the dusty plains to the east of our village awakened to the embracing morning light, my mother poured a gourd of water over my bowed head and then began the parting and braiding. She didn't finish my braids until the lone tree at the top of the hill to the west had become a black and gold warrior that preened against the evening sky and saluted the ocean as it submerged the sun.

I remember the melody of my mother's voice singing an ancient song or speaking the stories of our ancestors to help the time pass more quickly. I miss her voice, and I miss the feel of her fingers flying through my hair and the gentle tugs on my scalp as she made my hair beautiful. She had learned to make fancy braids from my grandmother, who had learned from my great-grandmother, who had learned from my great-great-grandmother, all the way back to the beginning of time, it seemed.

My mother could braid in straight rows like kernels on ears of corn or in spirals like snail shells. She could braid circles inside circles, and squares inside squares. But my favorite were the rows that curved up and down along my head, like the waves on the ocean I loved. My mother often wove in beads and attached several to the end of each braid. When I was a little girl, I liked to skip and jump just to hear them click and feel them bounce against my neck. As I grew up, I would sometimes toss my head to relive that memory. My favorite beads were red, and that never changed. And in my heart, I guarded the secret to the woman I am, and that never changed.

3

Family Stories

Though four inches taller than my mother, I felt small next to her. We sat on the sofa, the box between us. Mom leaned back into the pillows, her slender hands delicate, their joints and veins forming intricate angles and planes in her translucent skin. The woman at my side was much more than my mother: She was a woman of ancient times passing on a legacy for future generations. As she spoke, I reached for her hand and held on to it.

"Always remember — you're a Madison. You come from African slaves and a president," she said. Her thin, high-pitched voice resonated through my living room, repeating the words I had heard so many times, each word lingering with a hint of a Texan drawl.

"Exactly thirty years ago," she said, "when my daddy was very ill, he made me the *griotte*. It tired him out, but over three days, he told me all the stories, the ones passed down to him and the ones about his own life. Then, at the end of the third day, he reminded me that our history goes well beyond America's boundaries. What we believe in and what is important to us come from the vastly different beliefs and values people hold in Europe and Africa. And this," she said, searching my eyes, "is very important, Bettye: Each *griot* in our family has to understand that the Other Madisons might struggle sometimes to know how to live our lives, but when we share our stories, we build a sense of togetherness, and we learn who we are."

Though many in our family have heard we descend from President Madison and his slaves, only the *griots* know the full account of our ancestors, white and black, in America. Gramps had told me many stories, but the detailed history was Mom's responsibility to convey to me when I became the next *griotte*. That night, I understood for the first time why some of the details of our family history were passed only from the *griot* of one generation to that of the next. Not only were some of the stories intimate, but this tradition safeguarded their accuracy, truth, and longevity. I sank into the sofa with my mother and listened with a new awareness of the significance of her words and what they meant to me. She began.

"When your uncle John, your uncle Mack, and I were children, Daddy would call us to his library. He would spin the big globe that sat on his desk, then stop it to point to different continents and countries and teach us about ancestors who had lived there. When Daddy spun the globe and asked us, 'Who was the first African in our family to come to America?' John would stop the spinning globe, touch the outline of Africa, and shout, 'Mandy! She was kidnapped in West Africa and put on a slave ship.'

"Then Mack would butt in. 'They put chains on her, branded her arm, and shoved her down to the bottom of the ship!'

"'I was going to say that,' John always complained." Mom smiled with the memory of the childhood rivalry between her now-deceased brothers.

"The learning became a game. Over time, the boys lost interest, but I didn't. I can still see Africa and the different-colored shapes of countries on that old globe. Africa wasn't divided into countries when Mandy lived there. There were kingdoms, some big, some small, but I can't for the life of me remember the name of Mandy's. I only know that it was about where Ghana is now. Where she lived was probably once part of the ancient kingdom of Ghana. Daddy also told us where her village was, but I've forgotten that too. Years ago, I asked my brothers, but they didn't remember either."

I was disappointed that no one in my family could recall where Mandy had been captured. My mother explained that her memory was beginning to fail. Although she had forgotten some details, other members of the family had intentionally forced the knowledge from their minds. Such valuable information had survived more than 250 years, only to be lost in one or two generations. We could no longer place our ancestors and our cultural heritage in the world that existed before the transatlantic slave trade displaced and redefined us. Writing down the stories would save details from being lost in the recesses of aging memories.

In order to find evidence that our family had lived on the American landscape for centuries, my grandfather and, later, my mother sought out and saved what documents and photographs they could. Yet history had tried to deny our existence. Faced with such erasures and with lost or rejected memories, I could foresee losing everything, fact by fact, until, finally, there was silence — nothing to know, nothing to say. With the box now in my hands, I prayed for the ability to prevent the loss of any more information, no matter how unsettling or seemingly trivial.

"John Maddison — that's spelled with two *d*'s," Mom said, beginning the saga of the Other Madisons, "was our first white ancestor in America. His son, John Jr., kept that spelling, but his grandsons, John the third, Henry, and Ambrose, spelled it with one *d*.

"John Sr. was a poor English boy with big plans. Shrewd too. He became skilled at making boats and saved up his money to buy passage to America. He had learned about the headright system. It began in Virginia as a way to deal with the labor shortage. Growing tobacco required lots of land and lots of people to work it. The plan granted fifty acres for each indentured servant brought from abroad. For a specified number of years, seven or so, these poverty-stricken men, and sometimes women and children, worked to repay the price of their transportation to America. John wanted to take advantage of the opportunities in the colonies, so, in addition to saving up money for his own fare, he also put aside around seventy-two British pounds, enough to bring twelve workers.

"When he arrived in Virginia in 1653, he received six hundred acres, not bad for a young man who had next to nothing back in England. Most likely, John kept a few of the indentured servants to work his own farm, but to get ready cash, he sold some of the contracts to other planters."

As my mother retold John's story, I pictured a young white man wearing a vest, fraying trousers, a crumpled cap, and a determined look on his face. In the end, President Madison's great-great-grandfather might not have become as rich as he had dreamed of being, but he owned land, and landowners reaped respect and financial security. Selling and trading human beings for financial gain would continue in his family for generations to come. Those who profited from the headright system laid the groundwork for a way of life that relied on the enslavement of millions of black men, women, and children for nearly two and a half centuries.

"Mandy was our first black ancestor stolen from Africa. That happened sometime in the middle of the eighteenth century," Mom said. "She was the mother of our family."

I remembered listening to these tales as I stood beside my mother's sewing machine and wondering what Mandy looked like. Each time I heard her story, I loved Mandy more. My heart broke for her. She had come to America alone. She didn't know anybody, and everything was different. Mandy had no idea what was going to happen next or what she was supposed to do, and she couldn't ask; nobody spoke her language. But she learned how to pick cotton and tobacco and to speak English.

I wondered how her voice sounded. Over the years since then, I began to feel so close to her that I sometimes imagined her speaking to me. And I envisioned a dark-skinned woman growing old laboring in the fields of Virginia, her muscles withering and weakening. I saw Mandy become a slave.

"Mandy and her master, President Madison's father, had a daughter, Coreen," Mom went on. "She was the first African-American Madison and the second of our family's *griots*." As a child, I'd identified with Coreen because my coffee-with-plenty-of-cream skin was only a little browner than I envisioned hers had been. She cooked meals and baked in the Madison kitchen. According to our family history, apple pie was her specialty.

"James Madison Jr., the future president, saw Coreen walking back and forth between the kitchen and the mansion. And he wanted her," Mom stated simply. "As soon as she became pregnant with his child, she began to worry she would not be allowed to keep the baby for more than a few years. Coreen gave birth to a boy. She named him Jim. Raising him, Coreen lived in constant dread that he would be taken from her. She had heard of family members, even mothers and infants, being sold and separated by hundreds of miles, never to see each other again.

"In Jim's teenage years, Coreen's fear became reality. He had been born around 1792. A few weeks after his birth, Dolley's sister-in-law died, leaving two daughters — Susan, a toddler, and Victoria, an infant. Dolley agreed to take care of them. When the children arrived at Montpelier, she assigned Coreen to be Victoria's wet nurse. Coreen nursed Victoria on one breast and Jim on the other. The two children became inseparable.

"Many plantation owners believed that black people lacked the ability to read, write, or 'figger.' The slaves knew this was not true. As Jim grew up, he hid behind the door and listened in on Victoria and Susan's lessons. His father saw him hiding there but did nothing. Allowing Jim to learn," Mom speculated, "was Madison's way of showing love for his son.

"When Victoria was twelve, Dolley told her she could no longer be around boys, especially slave boys, like Jim.

"In 1809, when Madison became president, he brought Coreen and Jim to Washington. Dolley directed the other house slaves to make sure her niece was not around when Jim was working. But Victoria was hardheaded; she hid in the armoires in the Madison family bedrooms, where they shared their deepest thoughts and feelings. It didn't take long for Jim and Victoria to fall in love.

"One of the maids found out and suggested they stay away from each other. Jim was worried the maid would report them to Dolley, so he went to his mother. She burst into tears when he disclosed how he felt about Victoria. Coreen knew her boy could be sold or killed. She persuaded the steward to let Jim work in the kitchen, where it would be easier to keep the young lovers apart. But Victoria followed him there. One of the chefs, a slave, warned her that if she didn't stop sneaking into the kitchen, he would have to tell the mistress.

"In 1812," Mom continued, "the United States declared war on Great Britain, and on August twenty-fourth, 1814, British soldiers and slaves who had been freed and recruited to fight with them advanced on Washington. Dolley told Jim to save the American flag. He folded it, secured it under his shirt, then ran to hide in the woods with the other slaves.

"Years after the war," Mom said, "Jim told his children how worried he was that lightning and flames from the burning city would reveal his hiding place.

"In December 1814, two months after Britain and the United States negotiated a peace treaty, Dolley gave a party to celebrate. She ordered the male slaves, including Jim, to stand along the walls holding rushlights.

"A decade later, he told his children how sore his arms became, but he would never forget the music, the dancing, and the ladies and gentlemen in their finery. Standing there like a statue, he must have looked like the hero he truly was.

"A few days after the party, Victoria sneaked into the kitchen again. The chef promptly carried out his threat. Dolley was furious. She had assumed that Victoria and Jim knew better than to fall in love. Dolley arranged to sell Jim immediately, and the president made only weak objections.

"Just before Jim stepped onto the wagon that would take him

away, Coreen held him tight and wept bitter tears. Her only hope was that the Madison name might serve as a tool to help them find each other someday."

I knew what was next. I could hear Coreen whisper to her son: "Always remember — you're a Madison."

"Jim was sold twice," Mom went on. "The first time, he was sold to a nearby plantation, but Victoria drove herself there on a wagon to see him. The plantation owner sent her back and informed Dolley, who begged the new owner to sell Jim to someone far away.

"He ended up in Tennessee and never saw Victoria or Coreen again. But he remembered he was a Madison."

Jim, I would learn when I tried to find him, was one of the countless slaves who was not valued by those who created America's written record. Much of his story is lost to history.

There were more stories for my mother to tell me, but the hour had grown late, and Mom and I were tired. I accompanied her upstairs, kissed her good night, then returned to the living room. I have always been very private and reserved by nature, but I knew the source of my reluctance to accept this new role was about something more. For my mother, being the *griotte* meant being proud of descending from a U.S. president. She was not ashamed of having slaves in her family tree, yet, although most black Americans had enslaved ancestors in their genealogies, few had presidents. And she believed that President Madison was a great and good man.

I did not question Madison's greatness, but I did question his goodness. James Jr. inherited more than fortune and power. He also inherited the southern way of thinking and behaving. Madison was a Founding Father who was reputed to have been kind to the human beings listed among his possessions, but I knew that he, like his father and many other plantation owners, sexually assaulted or coerced the women he owned. In my eyes, the *griots* before me had glossed over the less than admirable behavior that had given James Madison Jr. a place of honor in my family tree.

I had to remind myself that Mom's view of the world was very different from mine. In fact, hers was only moderately different from that of women before the Civil War. Even white women from well-to-do families were supposed to be subservient to their husbands. Years earlier, I'd cringed when she told me she had attended college to study home economics in order to become a good wife. She had been working as a nutritionist when she met my father. After they married, she stayed home to take care of me and then my brother. Once we were off to school, she taught home ec in one of Oakland's public high schools, but Mom's raison d'être was tending to her home, her children, and her husband.

Mom entered womanhood during the Jim Crow era, when the effects of slavery were still in full force and when most black women were afforded little opportunity to become something other than a maid in someone else's house and a homemaker in her own. Mom was one of the relatively few black women able to obtain a college education and have a professional career, but she remained comfortable with some of the long-held beliefs on the limitations of being a woman.

I, however, had entered womanhood during the turbulent sixties, when America's young women challenged the status quo. We attended anti–Vietnam War rallies, burned our bras, and took The Pill, eager to step beyond the traditional roles for American women. We demanded to be taken seriously, to be respected, to be independent, and to be counted.

The moment my mother entered the kitchen the next morning,

I asked, "Do you remember how you tortured me when I was a little girl?"

Mom looked bright and energetic in her purple pantsuit and matching shawl. The sunlight glowing beyond the window lit up wispy filaments of her silvery hair. I was still in my faded nightgown. My graying brown hair lay tangled and flat against my head. I had not slept well, worried about what I would do with the box, the most important item my family owned. I was unsure whether I should store the heirlooms safely away so that they would not be lost again or share them whenever and wherever I could in order to bring to light all they represented.

"Tortured you?" Mom asked, an amused expression on her face as she sat down across from me at the kitchen table.

"Yes. You kept me imprisoned beside your sewing machine for hours at a time."

"Oh, that. Well, the results were worth the trouble. At the recital, you played the piano nicely, and you looked adorable. No one had the slightest idea what I had to go through."

"What *you* had to go through?"

"Yes, Dolly, what *I* had to go through. You would practice your recital piece without much fuss; just a little reminding did the trick. But when it came to the dress, you were too much."

"The only good part, as far as I was concerned, was that I got to hear the stories."

"I understood how you felt about the stories. When I was a child, I would follow Daddy when he went into his library to get some reading done or correct school papers. I knew he wouldn't turn me away if I asked to hear our stories."

Mom started from where she had left off the previous night. She told me about Jim and the generations of his descendants, including her own. By that night, I thought she was done with the handing-down. I thought I had heard all the stories. I was surprised when she said, "I have something to tell you before I go back to Oakland tomorrow. I've never told you before. Your father is the only person I've ever mentioned it to.

"Your grandmother never sewed for me, and she hated to cook. She thought sewing and cooking wasted her time. Always said she had better things to do, like prepare lessons for her students or mark papers. Every evening before she sat down to do that, she checked my homework and my brothers' too. A smudge or a single mistake meant trouble. And every morning, she inspected our clothes for the slightest wrinkle and our hair for a crooked part because, as she reminded us each and every day, we represented the whole colored race. So much was at stake, she said over and over. The white folks in town were doing everything they could to hold us back, to make us feel we didn't deserve to be free. Slavery ended some fifty-five years before I was born, but what those people really wanted was to turn back the clock and make us slaves again. It seemed like Jim Crow — 'Ol' Jim,' Daddy called him — lived right in our house."

"I knew Grandmuddy was strict, but I didn't realize she was that extreme."

"You haven't heard the half of it," Mom continued. "Bringing up John, Mack, and me couldn't have been easy, especially for Mother. She loved us, I'm sure, but sometimes it seemed she was trying to break us by enforcing a list of rules that was as long as her arm. She was hardest on me, claiming I was the most disobedient, and she was afraid to spank the boys. John's nose would bleed, and he had some kind of allergy that made sores open up on his legs every spring and summer. And Mack had a hernia. The doctors said to keep him still—which was impossible. So John and Mack got away with murder. I even got punished for things they had done.

"The worst offenses were using bad language and telling a lie. Even *lie* was a bad word. We were supposed to say so-and-so 'told a story.' When we said a bad word or got caught in a 'story,' Mother would take a bar of soap and scrub out our mouths.

"When I was the one caught, I got more than a mouthful of soap. Mother would push me face-down on the floor, sit on my back, and beat my behind with a belt. The more I screamed, the harder she hit me, and she did it until she got tired. By then, my backside was swollen and covered with red welts. And I was angry."

"Did Gramps ever intervene?"

"Daddy was never around when Mother hit me like that. What hurt even more than the beatings was that when I told him, he did nothing. The only way I could express my anger and grief was to refuse to address her as Mother. And I wouldn't even look at Daddy. But also, I think, to be honest with both you and myself, I was afraid I would see indifference in Daddy's eyes.

"I cannot justify her abusive actions to anyone, especially myself. I can only suppose Mother loved me in her way and was trying her best to direct me. I was a girl, so maybe she singled me out for punishment because she was afraid some man might try to hurt me if I didn't follow her rules carefully. Maybe Daddy understood her reasons and accepted her methods. A little misstep on my part could have cost everything they worked for and put me in danger. They were only one generation removed from being slaves themselves. They knew what had happened to many slave women. But John, Mack, and I were just kids. We had never seen a slave.

We didn't understand what it meant to be somebody's property. We didn't know what that meant for the women.

"What happened to Mandy was much worse than what happened to me. I tried to be strong like her. And Daddy had told me I was President Madison's great-great-great-granddaughter. I was proud of that. Even Mother couldn't beat that pride out of me. Mandy and the president got me through those beatings."

Mom never forgave my grandmother, and she never learned why my grandfather did not protect her. She had not mentioned the beatings to me before, and the box held no evidence of the abuse she had suffered. That narrative was a personal pain she had finally decided to reveal to me. She wanted her story to be complete, but I knew she would never talk about the beatings again. I wanted to say something that would take away that long-lasting hurt, but nothing seemed adequate. Mom blinked away tears. The beatings and Mom's anger and pain would now be part of the family saga.

4

Footsteps

Over the next two years, I went through the box again and again. I studied maps of West Africa and researched the history of slavery in America. I read biographies of James and Dolley Madison and historical studies on the relationship between Thomas Jefferson and his slave Sally Hemings. And I tried to figure out the message the directive should impart to the current and future generations of African-American Madisons.

My mother, like the ancient storytellers of West Africa, retold the family history that had been passed from generation to generation faithfully and accurately. She added her own stories and messages but never challenged any part of the saga. However, unlike the ancients, who relied strictly on oral tradition, my mother embraced a new tradition started by my great-grandfather Mack. To make the stories tangible and support them, he had gathered up letters, documents, and photographs. He wanted his children and grandchildren to see their ancestors, read their words, and hold in their hands evidence of what they had accomplished once they were no longer in bondage.

I, the newest *griotte,* would be the first to write it all down, and I began to realize I would be the first to explore the discomforting parts of our story. I had many questions. Who were these slaves and slave owners I had heard so much about? How had they influenced who I was? I wanted my ancestors to become real to me. In addition to Mack, Emanuel, Jim, Coreen, and Mandy, there were other enslaved ancestors living in my mind. I knew other names — Shelby, Henry, Charles, Young, John, James, Giles, Manda, Elizabeth, Lily, Katie, and Toby. I needed to visualize all of them and understand their sorrows, joys, and passions. I needed to know how Mandy had survived the Middle Passage and life in bondage and how generations of her descendants had endured unrelenting, sometimes life-threatening, racism. And I had to try to understand how James Madison could own some one hundred human beings while knowing that the widespread institution of slavery spat on the moral principles underlying the nation he had helped create. To know my ancestors and find myself in their stories, I had to walk in their footsteps. Virginia, I decided, would be the best place to begin.

My first visit to Montpelier, where James Madison had lived for most of his life, was in June 1992. I packed a suitcase, grabbed a couple of notepads and a handful of pens, and headed to the place where Coreen uttered the words that had guided my family for eight generations. I landed in Richmond, rented a car, and drove straight to the Virginia State Library.

A librarian sat at a table a few feet behind the waist-high counter. Thin and primly attired, with tightly curled brown hair, she faced me but kept her eyes on the piles of file cards she was organizing in a metal box.

I stood there a few moments and then said, "Excuse me. I'm

looking for any documents and books you might have on Montpelier, President Madison's former —"

"I know what Montpelier is," the librarian said in a thick southern accent. She did not look up.

"I came down here from Boston to do some research on the plantation," I persisted.

Her back stiffened. Still without looking up, she said, "Really? Well, you'll have to come back some other time."

Although thirty years had passed since the civil rights movement, I had come of age in that era and knew that resentment and prejudices die hard. *This* is *the South*, I reminded myself. To this woman, I was an invader, and the counter was her fortress wall.

I heard a voice behind me. "Hello. I'm Alice. I have a minute or two to help you if you'd like." A second librarian, just as white and with a drawl that was just as thick as the first librarian's, stepped out from the stacks. The first librarian finally glanced up. She rolled her eyes.

Alice, in her late forties or early fifties, was pleasantly round and very blond. She wore black-rimmed cat's-eye glasses and a yellow shirtwaist dress.

"Yes, I'd like some help," I said. "My ancestors were slaves at Montpelier."

"Figured as much," the first librarian mumbled.

"Have a seat, please," Alice said, directing me to a thickly varnished table and chair near her scratched-up desk. She vanished into the stacks and returned several minutes later pushing a clattering cart loaded with books and binders.

"This should get you started," she said, unloading everything in front of me.

"Thank you," I replied, then hunkered down to peruse records on the Madisons and on slavery in Orange County and other parts of Virginia. Time passed quickly as I became engrossed in dusty church documents, portions of Dolley Madison's letters to her niece Anna, and narratives by slaves and former slaves.

I read accounts of thousands of acres of tobacco cultivated by tens of thousands of slaves. I examined photographs of mansions and slave shacks, deciphered personal letters written by public figures and everyday citizens, and stared at notices of slave auctions. The Orange County of the eighteenth and nineteenth centuries came alive. It was clear that life in pre–Civil War Virginia was either grand or shabby, filled with optimism or devoid of hope, comfortable or desperate, depending primarily on the color of one's skin. White citizens, no matter how destitute, had freedom, had never had to fear shackles, chains, and whips. Blacks with "free papers" had freedom as fragile as the documents themselves. For slaves, freedom was an improbable dream.

I was beginning to understand how slavery shaped the lives of my ancestors, both slaves and slaveholders. It spawned the vulnerability of the former and the ironic combination of dependence and power of the latter. Antebellum Virginia was a place where masters treated their slaves as chattel, controlling what they ate, when they worked, when they slept, where they could go, and whom they could love. Yet the powerful were dependent on the vulnerable to cook and clean, take care of their children, work their fields, bathe, dress, and groom them, and, most important, keep them wealthy and secure.

Though I had much more to learn, I left the library with a sense of accomplishment. I drove westward, about seventy-five miles, to Orange County. I had reserved a room in a bed-and-breakfast and looked forward to the antique canopy bed, handmade quilts, a loft filled with nineteenth-century toys, and the seven acres of secluded lawns and trees described in the brochure. But I worried how my hosts would respond to me. From time to time throughout my adulthood, I had surprised more than a few folks by allowing my professional credentials to precede the arrival of my brown face.

I had spoken on the telephone with a pleasant woman about the room, and when she offered to assist me with sightseeing or antiques shopping, I seized the opportunity to "warn" her I was black, explaining that the purpose of my visit was to research my enslaved ancestors.

"How exciting!" she exclaimed, but I continued to feel wary. Exploring more than landscape and history, I was venturing into a living culture where slavery had thrived for nearly three centuries and where Jim Crow had been in full force until a meager three decades before my visit.

I drove up the gravel driveway to a small Victorian home. Leaving my luggage in the trunk — just in case — I walked across the lawn. Before I could reach the door, it flew open, and a petite woman with a bouncy ponytail ran out.

"Hi! I'm Pat," she began, grabbing my hands without dropping her knitting needles and yarn, while shooing away the cat close on her heels. "You must be Bettye. I've looked forward to meeting you ever since you called. So, you're from Boston. That's neat — Boston, such an interesting history, really pretty in the fall, lots of codfish, lobster, and beans too, museums and colleges galore, I hear. Of course, my husband and I have hosted folks from all over the country, all over the world, really, like England and Japan. Last week, we had a couple from Sweden, or maybe it was Switzerland . . . someplace like that. You wouldn't think it, would you, yes, right here in this small town, right in this little B and B."

Dressed in a pink tank top and faded cut-off jeans, she could not have been more relaxed. Pat was a who-you-see-is-who-I-am kind of woman. I smiled, then retrieved my suitcase from the car.

The next morning, I was the lone guest in the dining room while her shy, stout husband, Bobby, prepared and served home-made sausages, butter-soaked grits, and chunks of fresh fruit. After eating more than I should have, I followed his directions to Montpelier.

I parked at the visitors' center and took the shuttle bus past an antiquated train station and then along narrow roads through a wooded landscape. When the bus rounded a curve, the mansion, a stately, coral-colored jewel set before a sweep of hovering trees, came into view. A lush green lawn draped down the slope in front of the grand house. I had not expected to see a steeplechase track in front.

The moment I set foot on the soil where my ancestors had walked some two hundred years earlier, I felt that I belonged there with them, that they, and this place, would help me become the *griotte* I wanted to be. I started out on a guided tour with a group of visitors from New York, Wisconsin, and Florida, but I asked so many questions about slavery on the plantation that the guide referred me to the research staff.

I gave a brief overview of my family history to one of the historians, a middle-aged man wearing an ill-fitting jacket and rumpled shirt. He nodded and said, "Come with me." I followed him down a narrow, dusty hallway to a small room cluttered with papers piled onto every table and chair. "Bettye," he said, "this is Lynne, our chief archaeologist. Tell her about your family."

Lynne Lewis had obviously spent many hours on her knees combing through the dirt at Montpelier. She was suntanned, and her jeans were stained with Virginia's red soil. When I repeated my story, Lynne seemed to understand why it had been imperative for me to come to the plantation. As I soon learned, my research was important not only to me but also to the story of Montpelier, which the staff had recently begun trying to piece together.

"Let me show you something special," she said. "You're the first outsider — please forgive the term — to see it. We archaeologists tend to be a bit secretive when we first uncover something. Then, when we're done with the excavation, we brag about what we found to everyone who'll listen."

Lynne took me to the back of the mansion and pointed out a path of bare soil, about three inches deep and sixteen inches wide. Scattered tufts of grass grew along the sides. The path led from the rear entry of the manor to an excavation site about seventy feet away. Lynne knelt down beside a blue tarp and folded it back to expose an irregular rectangle of bricks stacked at one end of what had been the foundation of a small building.

She stood up, shoved her hands into the pockets of her Levi's, and rocked back and forth on her heels. "This was the kitchen," she explained, "and that mound of bricks is what's left of the cooking hearth."

What I saw was more than a hole in the ground and a heap of bricks, and the path was the groove that had been worn into the earth by generations of cooks walking to and from the kitchen, serving thousands of meals to the Madisons and their many guests in the mansion. My great-great-great-great-grandmother Coreen had walked this path. I stepped into the furrow and traced the length of it, a tether — both tangible and symbolic — that had held her in bondage. My eyes teared. I had not expected to walk so literally in the footsteps of an ancestor.

Stories about Mandy's first days on the plantation had been told again and again. Mom's clearest remembrances of hearing them were from the 1920s. She was a student in elementary school back then; Gramps was a schoolteacher in his early forties. When he got ready to tell a story, Gramps, wearing baggy pants, suspenders, long johns — even in summer — and a sweat-stained straw hat, would first lean back in his rocking chair on the front porch or in the squeaky swivel chair at his desk. Next, he would set the scene, then introduce each character with an identifying facial expression, hand gesture, or voice. As the story unfolded, he might imitate various sounds — a clopping horse, the rushing wind, a babbling stream.

Mom would journey with Gramps wherever his voice took her. After describing a West African village, he told her that Coreen's mother, Mandy, had been captured on the coast of Ghana. As Gramps told Mom about the Middle Passage, he imitated the sound of a creaking ship, and when he said that Madison Sr. bought Mandy, Gramps drew a cube and explained that slaves often stood on a big block of wood in order to be sold at auctions.

As Gramps told it, when Madison took Mandy to the plantation, he assigned her to a small, remote section where cotton was grown to clothe the slaves. There, in the cotton field, she "attracted his attention because she was such a good worker." I had heard this part of our family's story many times, but as I became an adult, I had many questions about it: Did Madison send Mandy to the cotton field because it was out of the way, far from scrutiny? Did he watch her for a long time before he decided to satisfy his lust? Did he violate her again and again? Was he brutal?

In the stories passed down by generations of my family's historians, nothing was said about what happened between Mandy and the master. Instead, the stories skipped from Mandy being a good worker to how Coreen got her name. Gramps wanted Mandy and Coreen to be as real to his children as the family they could see and touch, but he did not want to talk about rape.

The naming story, told in what Gramps believed to be slave dialect, his "slave voice," goes like this:

After seeing the beautiful little girl who had grown inside her body, Mandy said to the midwife, "Dere's a big ol' tree on a hill back home. I wants ta name my baby after dat tree so she grow up strong an' so she know where her roots is."

"No," the midwife replied, "best choose a name from dis place."

"I'll name her fo' de tree but won't tell nobody . . . but what ta call her out loud?"

"Let's find somethin' real pretty. Now let me see . . . you know, Mandy, jus' yesterday I was pourin' some fresh cream inta de churn. Dat cream looked so smooth, made you wants ta reach right out an' dip your fingers in it. I 'member now 'cause de butterfat floatin' on top was only a bit lighter dan dis here baby girl. What you got, Mandy, is a cream-colored baby."

"Name her Cream-Colored?"

"Don't be silly. Maybe jus' Cream . . . or jus' Colored," the midwife teased.

"Fo' goodness sake!"

The midwife chuckled and said, "Well, if you wants somethin' a mite longer, how about Coreamy?"

"Oh," Mandy said, "you de one be silly."

"Me? Never! But all right. I's go'n get serious now. How you likes Coream or Coreen?"

"Coreen. I likes dat," Mandy replied. "Yes. Coreen, my little cream-colored baby . . . Coreen."

At first, Mandy carried Coreen in a sling on her back when she went to the cotton field. When the baby cried, Mandy stopped to nurse for a few minutes. Sawney, the overseer, also a slave, allowed this interruption from work because feeding infant slaves helped sustain his master's business. Later, when Coreen had grown too big for Mandy to carry, Coreen stayed behind in the quarters with the other young slave children, watched over by a community grandmother. When Coreen was eight, she joined Mandy in the cotton field.

Another story Gramps told in his slave voice describes Mandy teaching Coreen to pick cotton:

"First we got ta thin out dese here l'il bitty plants, give 'em room ta grow. Den we's go'n chop up one row, down de next. Go'n keep on till we gets ta de end. After dat, go'n turn right back 'round an' pull out all de weeds. We's go'n pull up one row, down de next, so de cotton be healthy. Den we go he'p out wit' de tobacco, or we might get ta sit back a bit an' watch de cotton get bigger an' bigger. But near de end of summertime, we go'n be right back out here in de bright sunshine, walkin' up one row, down de next, pluckin' cotton an' stuffin' it in de sacks, bale after bale, sun up, sun down, sun up, sun down . . ."

Three years after Coreen learned to pick cotton, the head cook chose the young slave to work in the cookhouse. It sat many yards beyond the mansion so that if the kitchen caught fire, as kitchens were prone to do, the flames would not spread to the Big House.

Inside the cookhouse was a large fireplace with two or three bake ovens where several fires could burn at once. This was the Madisons' kitchen, and, except for the years when she assisted the chef in the presidential mansion, Coreen would work here for the rest of her life.

When I came to visit, that kitchen no longer had walls, nor did it have a roof, but seeing where it once stood and recalling the stories I knew so well brought a tightness to my chest. Coreen had worked here, and close by, she had stood helpless while Dolley sent Jim away.

Later, walking through the Madison family cemetery and trying to read names and dates on weatherworn tombstones, I felt the tightness again. My slave-owning ancestors, the ones who had torn my enslaved family apart, were buried here.

The past two days had taken a toll on me, but I said to Lynne, "I really should see where my black ancestors are buried."

"The gravesites don't have any names or dates on them. Besides, I wouldn't recommend that you go there in those shorts and sandals," she cautioned. "There's lots of poison ivy."

Overwhelmed and tired after spending days in the place where three of my ancestors had been held in bondage — two of them raped and one sold — I was relieved not to go to the burial ground. I needed to get out of there. Some of my ancestors were slaveholders; others were their slaves. The former wielded all the power. The latter were allowed none. I had to claim both.

Next time, I promised myself.

5

Living History

The following morning, Lynne and I met on the front portico, then walked a short distance beyond the northwest corner of the mansion to a small dome supported by ten columns. Like every other structure on the plantation, it had been built by slaves.

"Beneath the miniature temple you see here," Lynne said, "is an ice cellar. It's a deep cavern lined with bricks. Back in the days before folks had refrigerators, they went out in the winter and cut big blocks of ice from frozen ponds. Then they put the blocks underground to preserve food year-round. This temple is decorative, but Madison also intended it to symbolize his thoughts on philosophy and politics." She told me that Madison had built it to resemble a small version of the ancient Roman Forum's Temple of Vesta. The sacred flame that burned inside the original temple represented the safety, security, and longevity of Rome; in the United States, that fire became a symbol of liberty, what George Washington referred to as "the sacred fire of liberty" in his 1789 inauguration speech, which Madison wrote.

As we headed back to the mansion, Lynne said, "There's someone I want you to meet—Carolyn French. I'll set it up. She is the great-great-great-granddaughter of James Barbour. He was big in Virginia history. During the early nineteenth century, Barbour was governor of the state, ambassador to England, and a U.S. senator. He owned a huge plantation not too far from here and probably had some goings-on with one of his slaves. Carolyn is black. Some years back, she and her husband, David, a famous medical doctor, bought the home that used to belong to Barbour's great-granddaughter Winifred and her husband, John Albert Brown."

The next day, I drove the short distance from Orange to Barboursville. I found the address Lynne had scribbled down for me and turned off the main road. The car's tires crunched on gravel as I steered along a driveway flanked by shrubs and trees. I knocked on the front door, and a woman with light tan skin, wavy gray hair, and large, expressive eyes opened it.

"Welcome, Bettye. I've been looking forward to your visit," she said and gave me a hug. Carolyn, whom I guessed to be in her late sixties, was open and down-to-earth, the kind of person my mother would have described as "someone who never met a stranger." When I told her that my middle name was Carolyn, she laughed, clapped her hands, and declared, "You must be all right!"

She led me on a tour of her house. The original part, built at least 130 years earlier, was a modest four-room home with a central hall. New additions included large, gleaming white rooms with cathedral ceilings and Palladian windows. There was also a swimming pool. Carolyn's home was a far cry from the shacks where, not far away, three generations of my enslaved ancestors had slept.

As we sat at her dining-room table, sipping lemonade and eating sandwiches cut into dainty triangles, I explained the purpose of my trip to Virginia, adding, "Part of my family story is provocative."

"Good! I *love* provocative," Carolyn assured me, looking impishly over the rims of her glasses. "My family has some juicy stories too. But before we get into that, tell me about you — just the basics, for now."

"Okay . . . I was born in Tucson, Arizona, because my father was stationed at Fort Huachuca during the war."

"Born in a desert. That makes you a dune bunny."

"No one ever told me that, but you must be right. After the war, we moved to Northern California, the Bay Area. I grew up there, so I prefer ocean to desert. My parents have retired now, but Daddy was a doctor."

"Mine too."

"And Mom was a teacher."

"Mine too!"

"My dad's father was a doctor."

"My mother's father was a doctor. That must have surprised quite a few white folks. Most of them didn't think we should be allowed to learn to read, let alone go to medical school. Looks like our grandfathers knew how to kick Jim Crow in the butt!"

I laughed. "That's for sure. Nobody was thinking about affirmative action back then."

"Hardly. But keep going. Your husband?'

"Lee's a doctor, a neuroanesthesiologist."

"That's a mouthful. David's just an old country GP."

"*Just* a GP?"

"I have to keep him in his place," Carolyn answered, chuckling, the impish look still in her eyes. I would later learn that she enjoyed teasing her accomplished husband, who, it turned out, was not a GP at all. Dr. David French was a noted pediatric thoracic surgeon and a professor of surgery at Howard University. He was also a civil rights activist who had taken care of marchers injured in the Bloody Sunday attempt on March 7, 1965, to cross the Edmund Pettus Bridge from Selma to Montgomery, Alabama.

The more Carolyn and I talked about our upbringings, parents, husbands, and passion for history, the more we realized how much we had in common. Both of us knew that our mixed ancestry had come about because a powerful white male had "visited" a vulnerable black woman.

After about an hour, the doorbell rang. "That must be Ann. I told her you were coming. She wants to meet you. You'll like Ann," Carolyn said as she stood up and walked toward the door.

A slender white woman with long brown hair tied at the nape of her neck entered the room.

"This is Ann Miller. She and Thomas Obed Madden Jr. wrote a book called *We Were Always Free.*"

"I read that!" I said. "I'm really pleased to get to meet you."

"Would you like to meet Mr. Madden?" Ann asked. "He lives not far from here and loves company."

"Of course!"

"Then I'll give him a call tomorrow."

Ann was an architectural historian, but her knowledge went well beyond architecture. She knew nearly everything about the history of Orange County: dates, names, personalities, adventures, misadventures, legends, and facts big and small.

As a pediatrician, I wondered how slaves, who had limited resources, took care of their infants and young children. One of the questions I asked was whether slave babies wore diapers.

"No," Ann replied. "In fact, some white babies didn't wear diapers either. It was quite smelly and messy back then, even in the grandest homes. They had a different standard of hygiene in those days."

Ann loved Orange County and often described past events as if they were happening now. For her, history lived side by side with the present, and her reverence for truth was paramount to every other consideration, and it was color-blind. "The Madisons enjoy a comfortable life," she said. "Their slaves take care of their every need. And it seems that Paul Jennings, Mr. Madison's personal slave, can anticipate his master's every want. Paul is indispensable to James."

Ann phoned T.O. — as Mr. Madden preferred to be called — the following day. He was an African-American descendant of indentured servants who'd been at Montpelier during the latter half of the eighteenth century. His book was the story of six generations of his ancestors, and that included Irish immigrants, African slaves, and their descendants.

His great-great-great-grandmother Mary Madden was a poor Irishwoman who had came to Spotsylvania County, Virginia, in the 1750s. With no money and no resources, she became a charge of the state. In 1758, Mary gave birth to a "bastard mulatto," Sarah. Mary could not pay the pauper's penalty for producing a child, so when Sarah was not quite two years old, she was taken from her mother and bound as an indentured servant to George Fraser in Fredericksburg. A few years later, she would begin service; her servitude would last until she reached the age of thirty-one.

It is not known whether Sarah ever saw her mother again, but in 1767, when Fraser failed to pay a debt, Colonel Madison, as James Madison Sr. was called by his admirers, took over Sarah's contract. She was free, and her indentured status protected her from enslavement, but this nine-year-old child was poor and black and therefore nothing more than currency.

En route to visiting Madden, distracted by dense forests of towering trees along the winding, hilly road, I drove almost a hundred miles in the wrong direction. As the sun went down, the way back seemed more winding and narrow. I found myself leaning forward and gripping the wheel. By the time I arrived at Madden's home, the sky was pitch-black. I was drenched by a sudden summer storm as I dashed from the car to the front door.

The rambling structure was a historic landmark, once a busy tavern owned and operated by Willis Madden, T.O.'s Irish-African great-grandfather, before and during the Civil War. A labyrinth of dim rooms with low ceilings and uneven floors, the house was cluttered with aging furniture, books, and photographs. The former tavern felt cloistered and protective.

T.O., with his fair skin, straight hair, and hazel eyes, could have passed for white, and he looked far too youthful to be almost ninety. He greeted me in long johns and baggy pants held up by suspenders, just like my grandfather wore. Earlier that day, I had spoken with one of T.O.'s grandsons, and he'd told me that T.O. had recently been taken to the hospital because a car had backed into his legs. I was surprised he was so spry.

"I told my grandkids I was fine, but they dragged me to the hospital anyway. Now I'm using this cane to make the doctor feel better," he explained with a playful wink.

His sense of humor, in addition to his attire, reminded me so much of Gramps that I felt at ease right away, and T.O. gave me a big hug, as if I were a long-lost daughter. I was excited to be in his presence. His ancestors and mine might have stood face to face the way we were doing now.

The rain pounded the roof like a cascade of pebbles, and the wind rattled the ancient windows, but T.O.'s voice held steady above the din. "I'd lived here most of my life," he said, "but I'd never thought about seeing what was in the attic. A few years ago, I got curious and climbed up. The air was real stuffy and hot, so I didn't stay long. Just as I was about to leave, I saw a beat-up leather chest over in a corner. I dusted off the chest and carried it downstairs to this here bench. It took me a few days to go through everything, but I found my great-great-grandmother Sarah's letters, bills, receipts, and account books. There were also property records that included 'one old red cow.' A few birth records too."

Sarah, a skilled laundress and seamstress by the time her indenture ended in 1789, had saved these items, now more than two centuries old. Her small trunk sat like a symbol of the fortitude and survival of not only the Maddens but also the Other Madisons and countless other African-American families.

T.O. and I sat at a rough wooden table beside the chest. It was covered with peeling rawhide, and the lid gaped open. Though I was within arm's length of it, I could not bring myself to touch the chest. I knew all that it represented. When Sarah learned that her third owner, Francis Madison, one of Colonel Madison's sons, planned to sell her indenture and those of her four children, she decided to defy the conditions of her service and leave the grounds of Prospect Hill, Francis Madison's plantation. She went to Fredericksburg to plead her case to a judge. The judge granted her request that she not be separated from her children, but when she returned to the plantation a few days later, Sarah discovered she was too late. Three of her four children were gone.

As T.O. told the story, Sarah's strength hovered in the tavern. Her trunk was intimate and sacred, bulging with tangible evidence of the life she had built for herself despite uncertainties, hardships, and tragedies. The chest, I felt, was the place where she had sequestered her sorrows. Though Sarah went on to have many more children, she probably never knew the fate of the three who had been given away under the legal norms of the time. They might have become free after serving out the terms of their indentures, or, if their indenture papers were lost or destroyed, the children might have become enslaved for life.

Just as Coreen had lost Jim, and Mary had lost Sarah, Sarah had lost Rachel, Violet, and David.

History, I realized that night in T.O.'s home, is not just facts and dates. It is how people place themselves in the world. It is enslaved people who, in every moment of their quiet, invisible lives, stole pieces of themselves from their masters in order to say, "I am." History is in names that could not or would not be written down. It is in thoughts, feelings, and memories. It is a proud elderly man living in his great-grandfather's tavern along the back roads of rural Virginia. History is a thirty-one-year-old mulatto woman making a living for herself and her children and believing evidence of her life to be worth saving. Sarah's bundles of yellowing letters and piles of papers with curling edges reminded me of the box now in my care. Her trunk held stories and messages, and T.O. was its spokesman. He, too, was a *griot*.

I put nearly a thousand miles on my rental car getting a sense of the land, its meandering roads, red soil, thick forests, picturesque vineyards, rolling tobacco fields, and vast mountains. Whenever I spotted something of interest, I pulled over and climbed out of the air-conditioned automobile. I breathed deeply, enjoying the warm, humid air heavy with the smells of pine and wildflowers, animals and manure, and listening to the sounds of trucks whizzing down asphalt, horses clopping on sod, chickens clucking, and pigs grunting in yards. I watched tractors roll across fields where generations of slaves had worked centuries earlier. I saw airplanes zoom overhead, their passengers oblivious to the old and new ways of life woven together in many dynamic ways on what, from so high above, must have seemed a static red-and-green mosaic.

Nights at the B and B, I walked barefoot across the cool lawn as hundreds of flickering fireflies tried to compete with the thousands of steadfast stars in the amethyst sky. The air was noisy with nighttime. I imagined Mandy, Coreen, and Jim walking on similar soft, cool grass, amazed by the sparkle of fireflies and stars and the crickets singing in the dark. I felt as though I were sharing a moment with my ancestors.

During the day, as I drove through small towns, explored shops that sold everything from fishing lures to quilts, rested on park benches, or sat in restaurants where I developed a craving for meatloaf doused with a sweet sauce, I watched the citizens of Virginia as they walked from place to place, worked, ate, and chatted together. In public locations, Virginians, white and black, practiced the art of friendly aloofness. Their voices were cheerful, their facial expressions pleasant, but after a smile and brief eye contact, they looked away, back to business, back to their thoughts. The Virginians I watched were never rushed or rude, as people so often were in the North, but it seemed to me that in Virginia, after the hellos and how-are-yous, a wall of social hierarchy went up. Men in business suits and women wearing pearls seemed to maintain a tolerant disregard of anyone in overalls or an apron.

In private homes, however, southern hospitality was no myth. I joined an inclusive kindred. Some of us were white, others black. Some of our ancestors had been historic figures; some had been indentured servants or slaves. Some were seasoned archaeologists, genealogists, or historians, and some — like me — were new to it all. But we all shared the same thirst to know Virginia's past and how we fit into our land of the free and home of the brave. I realized that I belonged among these adventurers, and they took care to include me.

Years ago, a black southerner told me that in the South, whites do not care how close blacks get as long as they do not get too high, but in the North, she said, whites do not care how high blacks get as long as they do not get too close. So far in Virginia, with the exception of my brief encounter with the first librarian, I had not had to confront racism. But that would change.

6

Destination Jim Crow

O f course, I had been to the South before. Forty-four years earlier, in the summer of 1948, my mother decided that a trip to see my grandparents was long overdue. I was going to ride in a train for the first time!

Mom's decision to set out from Northern California and take me to Texas was not made lightly. The South held too many persistent memories and haunting associations. Since her departure from there in her early twenties, she'd gone back from time to time, but she had had her way: Her child would not grow up in a place where black people were lynched.

One of the stories my mom told me again and again, and one I, too, would pass down, was about her own first train ride. Mom had been born in 1918 on the "colored" side of Elgin, a small town near Austin, Texas. Ruby and her two brothers, John Jr. and Mack, grew up in a two-bedroom, white-frame house. Fewer than twenty feet separated the back door from the back fence. At least a dozen chickens strutted about the yard, and three huge hogs lolled beside a weather-beaten feeding trough. The Blackland Prairie soil was always velvety and dusty, even minutes after a rain, and littered with weeds and chicken feed. Hoes, shovels, picks, rakes, pitchforks, and hatchets leaned against the house, the fence, and the trough. A railroad track stretched behind the back fence. Past the front porch, two rows of nearly identical homes, separated from each other by narrow strips of dry grass, flanked the dirt road.

Located less than a mile away from the colored neighborhood, Elgin Union Depot was a switching point for two railroad lines that crisscrossed and linked the southern states. The Missouri-Kansas-Texas Railway ran north and south and exchanged cars with the Southern Pacific line, running east and west. Yanking and pounding the heavy steel links and pins, the station crew uncoupled cars from one train and then, in a matter of minutes, coupled them to a different locomotive heading out. About twenty trains a day stopped at Elgin, and many of these switched cars. The little country town was truly an American crossroads.

From their backyard, my mother and my uncles could watch the MKT line pass by. Whenever a train thundered down the track and blew the smell of hog-corn-chicken-slop-dust through all five rooms of their home, the three curly-haired, golden-brown siblings would shout, "There goes Katy!"

Their father, my Gramps, lived in Elgin for most of his life, but his dreams encompassed the globe. In his early childhood, he fell in love with trains, sensing, even then, that the massive steam engines and railroad cars could carry him wherever his curiosity and imagination led him. For him, trains were both physical and symbolic links to the alluring world beyond a small southern town. He subscribed to *National Geographic* magazine and saved every issue for nearly thirty years. A teacher in a one-room schoolhouse, he taught the colored children, including his own, not only reading, writing, arithmetic, and American history, but also world geography and culture. He agreed with the intellectual W.E.B. Du Bois that "it is the trained, living human soul, cultivated and strengthened by long study and thought, that breathes the real breath of life into boys and girls and makes them human, whether they be black or white, Greek, Russian or American." But Gramps also agreed with pragmatist Booker T. Washington's statements that one should "dignify and glorify common labor. It is at the bottom of life that we must begin, not at the top" and that "no race can prosper till it learns that there is as much dignity in tilling a field as in writing a poem." Therefore, Gramps's older students learned planting and harvesting, selling and accounting, in addition to their academic subjects. A scholar who dressed like a farmer, Gramps loaned his students books and magazines from the bulging shelves in his library and farm tools from his cluttered backyard.

When Ruby, John Jr., and Mack were toddlers, Gramps taught them to identify colors by practicing with the painted freight cars. Later, he used the cars to teach them the alphabet and then words — *Texas, Kansas, Southern, Railroad*. He described the attributes of locations he had seen or read about — the steep hills and rolling fog of San Francisco, the excitement and energy of Harlem, the music and riverboats of New Orleans — places where a train could take them someday.

Still, Gramps cherished Elgin. The colored side of town was a close-knit community. Women in families with abundant food "fixed up" plates and carried them down the road to those in want. Men with cars gave rides to anyone needing to travel farther than walking distance because buses were few, and seldom was there room at the back.

Every adult was responsible for every child and tried to protect them all from the hatred that seeped in from the other side of the railroad track, and every child was answerable to every grownup. Failure to stay close to home was sure to provoke adult wrath. Home was the only place where the older members of the colored community could shield the younger ones from white children who, with impunity, called them names, yanked their hair, pelted them with rocks, and smeared them with mud or animal feces. Anywhere in the South, including Elgin, colored teenagers and young men who ventured too far from home could be arrested and imprisoned or shackled to chain gangs, often for five years or longer, for misdemeanors as minor as "laziness" or "impudence." My grandfather learned this lesson well. When Gramps was fifteen, two policemen arrested him for sitting on the step behind the train station. He stayed in jail for two nights. He was lucky.

But Gramps knew that the luck of his youth would wane and that he could not keep his children safe forever. He saw that life in Elgin, on either side of the track, was restrictive. Social and educational options depended on the churches (one white, one black) and on the schools (one white, one black). And employment opportunities were limited, especially for the colored citizens. Men were sharecroppers or laborers in the cotton-oil mills, the slaughterhouse, or the brick factories. Women were maids or took in laundry. Everyone picked cotton. Gramps picked cotton throughout his life. Even after becoming a teacher, he returned to the cotton field to work side by side with his children when they wanted to earn spending money.

Gramps wanted more for his sons and daughter than Elgin could offer and was determined to send all three to college, but he realized that once they graduated, his children would probably not return. He would stay in Elgin because he was dedicated to the community, especially his students. Moreover, his parents lived just a few miles down the same railroad track. Gramps loved Elgin, but he loved his children more. And each, with his blessings, would find a way out.

For Ruby, my mother, the way out was to study home econom-ics. Well before setting off for Prairie View College, she promised herself she would never be anyone's mammy. She would never cook in someone else's kitchen, scrub someone else's floors, or raise someone else's children. In her early teens, as she did her homework on the front porch, she saw neighborhood women trudge home dressed in maids' uniforms, exhausted after a day of housework and childcare on the other side of town. In Ruby's eyes, the women were trapped. She mapped out a different life. Like her aunt Ruth, the younger of Gramps's two sisters, she would study home economics in college and become a home ec teacher in a big-city high school. Ruby would choose a college-educated husband and rear their children far from the South. She would or-der the most fashionable clothes from the Sears, Roebuck catalog: sleek silk dresses, suits with fitted waists and squared shoulders, felt hats, sheer stockings, and two-toned spectator pumps. Best of all, Ruby was certain, she would be a passenger in one of the Pull-man cars that rolled by her backyard.

Working as a maid, she felt, was a fate, but teaching home ec — scientific and focused on efficiency and propriety — was a ca-reer. Her first teaching position was in a lumber-mill town in Texas mere steps from Arkansas. Four years later, she found work in Kansas City, Missouri. Ruby was finally out of Texas. Though once a slave state, Missouri was farther north than she had ever lived. When she was twenty-four, she went to St. Louis to visit her brother John, a physician in training at Homer G. Phillips Hospital, and her aunt Estelle, the superintendent of nurses there. By divine providence, Ruby met a man who could ensure she would never have to live in the South again.

For the first ten years of his life, Clay Morgan Wilson III enjoyed a comfortable existence in Shreveport, Louisiana. His father, Dr. William Douglas Wilson, had a successful medical practice, and his mother, the former Ellen Guesnon, taught music in their home, an elegant Victorian with wide porches and awnings. Clay had two "tolerable" older brothers, W.D. and Eddie.

One hot summer afternoon in 1921, a group of white boys chased Clay's brothers the several blocks from the local ice cream parlor to their home. Gasping for breath, W.D. and Eddie leaped up the steps and ran into the house. As soon as the door slammed behind them, their father ran out onto the porch wielding a shotgun and threatened to shoot anyone who dared to bother his sons again. The white youths ran away, but when Dr. Wilson calmed down, he remembered that Louisiana was not a place for any black person to confront any white person for any reason. Moments later, employing his usual flat humor and still holding the shotgun, he turned to his family and said: "Now it's just about time to leave here." He rented a freight car and loaded it with nearly everything they owned, including his convertible-top automobile. In less than a month, they were on their way to Oakland, California.

Years later, my father loved to recount the moment that changed his life. It wasn't the day he and his family fled Louisiana; it was the day he first saw my mother.

During his medical internship, he was in a hospital elevator and had just pressed the button to go to the fifth floor — obstetrics

—when "a young woman, a real looker," stepped on. She pressed the button to go to the third floor, but the elevator carried its two occupants nonstop to the fifth. Obstetrics had priority. On the ride up, peripheral vision served Dr. Clay Wilson well. When he got off the elevator, he sought out a few nurses and doctors, described the woman on the elevator, right down to the pearls clipped to her delicate earlobes, and asked if anyone knew her. Everyone laughed, but within ten minutes, he learned that the fashionable young woman was the niece of Estelle Massey and the sister of Dr. John Madison, one of his own colleagues. She was a nutritionist and home economics teacher visiting from Texas, and, most important, the lady was single. And when he heard her name, he recalled the color of her lips and smiled at how well it suited her. Ruby.

My father used a substantial portion of his meager intern's salary to buy a 1940 maroon-colored Pontiac. Now he was ready to court the poised, elegant, and intelligent graduate of Prairie View College. Little more than three months later, on August 16, 1942, Ruby and Clay were married. At the civil ceremony that summer afternoon, Ruby was serene in a pale blue suit, but the groom, who had hoped to look regal on his wedding day, sweated in a heavy woolen army uniform. He had received an "invitation" to join the armed services. Failure to comply meant he would be drafted as a private. If he accepted, he could serve as an officer in the medical corps, and that service would be considered equivalent to a residency. A few weeks after the wedding, the newlyweds departed for Fort Huachuca, Arizona, an army base for black soldiers. Except for an occasional visit, Ruby was away from Jim Crow for good.

Ruby had learned about racism early. From their childhood backyard, John Jr., the eldest and most serious-minded of the siblings, was the first to notice that the train engineers and caboose men were white and that the men who pumped handcars up and down the tracks were brown-skinned Mexicans. Not a single black man rode past. By the time she was four, Ruby, the shy one, realized they had not seen a black man on the railroad. Mack, the youngest and most energetic, was not far behind. When the siblings asked Gramps why there were no colored men on the trains, he explained, "Jim Crow — Ol' Jim — makes the rules. But," he added cryptically, "wait to see what you see when you go for a train ride."

Mom told me about her first train ride in such detail that I can recount it myself.

In 1923, when Mom was five, her mother — "Grandmuddy," as I later called her — loaded a basket with fried chicken, potato salad, and biscuits, then she and her three children dressed carefully for the trip and headed to the train station for a day-and-a-half ride to visit relatives in El Paso, Texas. Aware that the colored car was near the front of the train and that their clothes would become blackened with soot from the engine, all the colored passengers put on dark clothes — a brown cotton dress for Grandmuddy, a dark blue dress with white plastic buttons down the front for Ruby, and brown wool jackets and knickers for the boys. Everyone — whites in their colorful outfits on one side of the station, blacks in their somber-hued garments on the other — waited, perched on wooden benches, luggage all around.

The moment the train came into view, the travelers on either side of the station jumped up and smoothed out their clothing. Men collected the luggage. Women collected the children. As the train chugged to a stop, my mother and her brothers saw porters in handsome blue uniforms suddenly appear, as if from nowhere, and throng the station. They were colored! Colored men did work on the railroad! The porters rushed around, calling out to each other, assisting one passenger and then the next — all the whites before any blacks — to board the train, throwing sacks of letters into the mail car, and heaving suitcases and trunks into the baggage car. The three children stood and stared.

On board, Ruby, John, and Mack caught sight of the dining-car waiters. They were colored too! The grace of the waiters — polished gentlemen who rang hand chimes from car to car, announcing each meal — filled them with awe. Dressed in white jackets and slacks, the waiters glided, straight-backed and effortlessly, in rhythm with the music of the chimes and the hypnotic chug-hum sound and sway of the train.

The colored car was crowded. Day or night, men slept in nearly every row, their heads propped against the windows or hanging back over the seats, newspapers shielding their faces from the glare of naked light bulbs on the ceiling. As the men slept, Grandmuddy and the other women tried to control the children, many of whom, including Mack, were climbing over the seats and running, laughing and shouting, up and down the aisle.

When Ruby and her brothers peered into the coach car for white passengers, located just behind theirs, they discovered it was less crowded; the windows and floors were less sooty, and the upholstery was less shabby. But here, too, the siblings observed, most of the men were asleep, newspapers shielding their faces, and the women were trying to control raucous children.

While the train sped along, the adventurous trio explored the other passenger sections. The dining car had white linens and a lamp on each table, a mysterious curtain at one corner of the room, near the kitchen. In the club car, there was an ashtray on a stand beside each lounge chair and card table. And, finally, for well-to-do white passengers, there were the Pullman cars, each had a bright corridor with windows along one side, a row of steel doors on the other.

These fortresses far from the pandemonium of common folks were the domain of the Pullman porters. Wearing meticulous white cotton jackets and dark blue pants, the porters went from one private compartment to another, taking away shoes to be shined, delivering newspapers and magazines, bringing a glass of water, turning down bedding. Being a porter was a coveted job. They remained stoic and dignified when white passengers or supervisors called them "boy." Their voices were clear, their diction perfect, but softened by the southern cadence. "Ma'am, may I please assist you?"

As night approached, Ruby, John, and Mack entered the last Pullman and peeked through the open door of a stateroom. Later, my mother would remember every detail of what she saw: A white man sat on a high-backed, red mohair sofa; he wore a beige suit and a matching Homburg hat. A tall woman stood in the compartment, her back to the door. From one side of her close-fitting gray hat, an iridescent blue feather curved high over her head. She wore a pale blue suit and soft-looking gray gloves. Ruby noticed that the seams of the woman's sheer stockings were centered perfectly down her legs. Not even on Sunday mornings at the colored church in Elgin — where each woman in the congregation had spent at least an hour the night before washing and ironing her best outfit and adjusting the artificial flowers on her straw hat — had Ruby seen anyone so elegantly dressed.

Two blond children sat on either side of the man. The little girl, about five years old, wore a frilly pink dress prettier than any Ruby had ever seen. Black patent-leather Mary Jane shoes glittered when the girl joyfully kicked her feet up and down. Ruby looked at her own ankle-high boots, which, until this moment, she had loved. The brown leather was dull and had creased and cracked across her toes.

The girl's younger brother wore a cream-colored suit with short pants. His black shoes were smooth and polished, the laces tied just right. The white calf-high socks on his stubby legs were straight and even, but his exposed knees gave away his secret — he was no different from any other child. Covered with scrapes and bruises, his knees were just like those of the children in the colored car. The porter, unaware of the observers at the door, ignored the blemished knees and treated the scuffed-up little boy with the protocol and esteem set aside for all first-class passengers. He held a silver tray while the child ate vanilla ice cream from a crystal bowl.

The boy was the first one in the stateroom to see the spectators. He looked up from the ice cream, stuck his fingers into it, and slowly licked each finger, taunting Ruby and her brothers. Then he stuck out his tongue at them. Following the boy's gaze, the porter stiffened, scowled, and rapidly flicked his wrist, shooing the three trespassers away. John turned his back. Mack stuck out his tongue at the boy. Ruby — stunned that the porter, unlike the adults who had watched out for her in Elgin, had not come to their defense — cried.

When they returned home after the trip, the children told Gramps how the porter had treated them.

"I think," Gramps said, "he was just trying to hold on to his job. Porters can get fired for any reason or no reason whatsoever. They work from the crack of dawn until all hours of the night, and they don't get to see their families for days, even weeks, at a time, not until the run is over. But a job on a train is a good job, especially for a Negro. And keep this in mind: no matter what might happen to a colored man working on a train, he is much safer there than in any southern town, where Ol' Jim runs the show."

Gramps wanted to say more, but his children were too young. A few years later, Mom recalled, Gramps explained that a porter could glance briefly at a white woman while serving her on a train, but on or off a train, any colored man in any position would almost certainly be murdered by a mob if his glance lingered. As a child, Ruby was confounded and terrified. As an adult, she understood but was no less horrified. Then, riding in a car one stifling summer day in Texas, she saw a dark-skinned man hanging from a tree, the edge of a cotton field glaring below his feet like a knife. Ruby did not know if he was there because he had dared to look at a white woman a moment too long, but it was impossible to forget the harsh angle of his neck, the branch, the rope.

When my father left Fort Huachuca to serve as a lieutenant in the medical corps stationed in the Philippines, my mother and I traveled to Pasadena and stayed there for a few weeks with his mother and sister. We then settled in Oakland, where my father had grown up and where he and my mother had decided to raise their children. In California, there were no filthy public toilets or rusty drinking fountains under signs that said COLORED. The schools were integrated, and no one picked cotton to get by. By the 1940s, California was home to a large number of black transplants from the South determined to make a better life for their children. The Negro citizens of Oakland, like those of Elgin, felt a strong sense of community. The adults kept a watchful eye and pooled their energy, resources, and talents to support and guide the youth. Whenever racial slurs were thrown our way, adults told us that the slurs showed ignorance and that we should not pay any attention. After local grocery stores refused to hire colored students for the summer, proprietors of Negro-owned shops created positions for us. And any youth unjustly arrested was bailed out of jail with funds collected in the black churches.

From the moment she got there, my mother thrived in California. She was a southern girl who felt safer and more fulfilled in the North, which for her was anywhere that was not the South, a place that haunted her with its shameless racism, demeaning segregation, and sanctioned murder.

But we had roots in the South, roots we would not turn away from no matter how painful the memories or how cruel the history. It was where the stories about our ancestors had taken shape,

My mother, Ruby Laura Madison Wilson, and me, circa 1948

and perhaps she wanted to share with me the indelible images of her first train ride, when she, too, was five years old. My first train ride would become, like Mom's had for her, the most vivid memory of my childhood. And it would reveal that escaping from the mores of the South was not easy.

My first train ride was on the Southern Pacific Railroad. I had just turned five years old when my mother and I rode from Berkeley, California, to Navasota, Texas. My grandparents had lived there since 1940, when Gramps reluctantly left Elgin to become the principal of a much larger school.

When Mom and I arrived at the station in Berkeley, I noticed that the porters were colored, just like me. I was not surprised. One of my grandmother's brothers and several of my parents' friends were porters. I enjoyed watching a large group of them working together. They were strong, full of energy, very good-looking, and had huge smiles. They all wore snazzy dark blue uniforms embellished with gold braid and shiny brass buttons, and blue caps with SPR on the front. But each man used his hat — tilting it to one side or low on his brow or way on the back of his head — to make himself one of a kind. The caps never fell off, not even when the porters hurried around swinging luggage and hoisting boxes like they were dancing, the big showoffs.

Before the train arrived, the porters talked and laughed together, their suits converging to form patches of blue on the gray concrete platform. Then the train came into view, its whistle louder and louder as it approached. Billows of black smoke drifted upward and disappeared into the sky. The engine got so close that all I could see was the gigantic grate on the front. When the wheels screeched to a stop, I covered my ears. The blue suits swarmed.

My father had never worked at a station, but each summer during his college years, he was fifth chef—the dishwasher, really—on a train. The waiters and other chefs treated him like a son. Later, when he became a physician, he had a special fondness for black railroad workers and gave free care to those in need, so Mom and I received special treatment. The porters picked up our bags first and then escorted us to the train. Mom was dressed to the nines in a chic black hat, soft black gloves, shiny black high-heeled pumps, and a wide-shouldered sage-green suit. I had grown impatient as I watched her take forever to make sure the seams of her stockings were perfect.

Suddenly, I was high in the air, squealing, pigtails flying, skinny brown legs dangling below my new pink dress. Then, feet first, I landed softly at the doorway to a Pullman car. The porter who had lifted me bowed. Mom smiled. Other black passengers laughed, but several white passengers scowled. I now know that for them, it was insult enough that the cars were not segregated in California but nearly intolerable to see such a fuss over a brown-skinned child.

The entrance opened onto a long hallway that had a row of windows on one side. On the opposite side, one of the doors led to our compartment, where everything was gray and smelled of wheel grease and recently cleaned carpet. The door remained open, and as I stood on tiptoe to try out our very own stainless-steel water fountain, I glimpsed a tall colored man in the hallway. I knew he was a Pullman porter because he wore a starched white jacket with a high, tight collar and a bow tie. His pants were dark blue, the creases razor-sharp. His shoes and socks were black. The brass plate on his cap, Mom had told me, said PULLMAN PORTER. She had described his outfit perfectly.

"He remembers when your father worked in the kitchen," Mom informed me.

"Ma'am and little lady," he said as he stepped in and handed me a glass of apple juice from a silver tray, "you can call me Harold." I thought I saw a smile, but he did not say another word.

A whistle blew. The train groaned forward. The engine picked up speed. I ran back and forth, faster and faster, between the window in our compartment and the ones along the hall, a row of frames through which changing pictures zipped by as the train raced past towns and farms, through valleys, over rivers, and around hills. When the heavy steel door at each end of the car slammed shut, the metal-to-metal roar of wheels on tracks hushed down to a hum. Meanwhile, Pullman porters disappeared into and reappeared out of the compartments, ignoring the kaleidoscope of scenes beyond the windows.

Mom told me to close the door and sit down. I obeyed, the silent man observing from his post in the hallway. Harold's face was serious but kind. I could tell he was there to watch over me. I had no idea why.

That night, our train stopped in the middle of a desert. The sun had set but left behind a shimmering orange glow. The wide shadow of purple mountains vanished into the magenta sky that grew darker and darker. Except for a small, lit-up train station, the endless desert floor looked black and empty. Only two passengers boarded the train, a woman and her daughter, a girl about my age. Both had blue eyes and curly blond hair. The mother had on a neat white blouse and a gray pencil skirt. The child wore a blue dress with puffy sleeves and fancy stitches across the bodice. I noticed her outfit and thought that both our dresses were pretty. The lady spoke pleasantly to the porter, and the little girl looked up at him, grinned, and shouted, "Hey!"

The woman nodded to my mother, and her daughter gave me a big smile. I smiled back, hoping we would play together. Harold moved a little closer to our cabin doorway. Stepping into the corridor, I watched him escort the newcomers past our cabin and hold open the heavy steel door at the end of the narrow passage. He did not lose his dignified deportment as he looked back at me.

The next morning, I heard a light tap on our door. I took my time responding because I thought it was a waiter who had come to announce breakfast. And I was a picky eater. Mom urged me to hurry up and open the door. When I opened it, I saw the little girl hugging a doll with blue eyes and Shirley Temple curls like her own. Her mother looked at mine, seeking approval. My mother motioned for them to come in. Behind the woman, Harold watched.

"I'm Susan," the lady said.

"I'm Ruby," my mother replied

"My name is Mary," the girl said.

"Mine is Bettye," I said.

For the next two days, Mary and I were inseparable. Giggling nonstop, we played with dolls in our compartments and shared crayons and paper in hallways. Once, we sneaked behind my mother when she went to play canasta in the club car. The oblong space, where both black and white passengers laughed and mingled, was hazy with smoke, the air thick with the sharp, sweet smell of burning tobacco. The room was noisy. Ice tinkled in glasses; red or blue playing cards slapped down onto aluminum tabletops; voices murmured in conversations about destinations: "Out west," "Back east," "Up north," "Down south" . . .

At some stops, the porter hopped off the train to buy food for passengers from women holding covered baskets. Mary and I and our mothers ate roast beef or turkey sandwiches in our cabins. Meals in the dining car were "outrageously expensive," my mother told me. More outrageous, she said, was that when the train reached the South, colored diners would have to sit behind a curtain. I thought that eating behind a curtain sounded like fun. At home, my friends and I dragged toy furniture behind the drapes and played teatime with our dolls.

One evening, Mom surprised me with a treat. She had ordered fried chicken from the dining car and invited my new friend and her mother to join us in our compartment. A waiter came, balancing an enormous tray on one shoulder. Beside the sparkling glasses, shiny silver domes covering the plates reflected his smile and stretched it wide. When he lifted the domes, the paper ruffles on the drumsticks flabbergasted Mary, but my mother had already told me about decked-out chicken legs.

The next afternoon, as I was about to go meet my playmate, Harold appeared at our door. He looked over my head at my mother, a frown wrinkling his forehead. Worry flashed across Mom's face. Something had happened. The train was at a station that looked like all the others, but when the outside door opened, a blast of hot, humid air filled the car. We were in Texas.

I stepped forward to find Mary, but Mom called me back. "Why?" I whined and protested for a moment, but from the look on her face I knew I should not press her further. Harold glanced at me, then started down the passageway toward Mary's car. Mom closed the door to our compartment, and I played alone with my doll, pouting.

At the next stop, I went to the window to see who would get on.

My mother was close behind me. I pressed my forehead against the glass, and what I saw bewildered me: Mary stood on the platform between her mother and a pile of straw suitcases. They had reached their destination, I realized. Holding her pink plastic doll case and her white sweater, Mary searched the train windows. She found me, and we began to wave a tearful goodbye, the steel and glass separating us. As my mother wept for me, Harold stepped through the open cabin door, knelt down, and held me gently. In the years to come, I would grow to see that this childhood moment was my introduction to Ol' Jim himself. I lived in California; I didn't know there were places where friendships between black children and white children were forbidden. In 1990, when it came my time to be the family *griotte,* I would add this childhood memory to the stories to be told to future generations of Madisons. But on the railroad in 1948, the two mothers, in unrehearsed unison, lowered our waving arms. Then the train, slowly at first, carried me deeper into Texas.

7

The Dentist

Among Gramps's many contributions to our *griot* tradition was his pragmatic philosophy. Mom told me that when she and her brothers asked why there was a colored car and why it was the dirtiest car on the train, he answered with just one word: "Racism."

He told his children that experiencing racism was as much a part of them as their name. Mom remembered her surprise when he said, "I'm glad we colored people have to fight it." Gramps explained that he believed racism made the black community more cohesive and helped its people see what everyone deserved and should work toward. "And," he told the children, "when we stick to it, we get it. Maybe it's hard to believe sometimes, but we have the upper hand in the fight. Racists are scared, and racism is just another challenge, and challenges make us strong." He ended with this reminder: "White folks may not like it, but here we are, and here we'll be. They don't know it, but we know we are not victims. Racism is a feeble leftover of slavery."

Gramps was born free, but his father, Mack, had been born a

slave. In his lifetime, Mack had been threatened, humiliated, and sold, but he believed in himself, Mom told me. After emancipation, he worked as a sharecropper until he had enough money to buy land. For the first time in his life, he owned something. Mom remembered Gramps's words: "You children have never been slaves; neither have I. We own plenty of things, so surely we can put up with separate toilets and drinking fountains. We can't touch produce when we buy food. We can't try on clothes before we purchase them. We have to lower our eyes when a white person looks us in the face. And, if we're lucky, we sit at the back of the bus. So what? Those foolish things don't stop us from being who we are: Madisons. Americans."

No one believed in the family saying more than Gramps. No one was prouder of his ancestors, black and white. He was as proud that Mandy had survived the Middle Passage as he was that President Madison had helped found our nation. Without Mandy, we would not have our strength. Without Madison, he said, we would not have our name.

Time and again, Gramps told his children that America, conceived by Madison and the other Founding Fathers, would never have become what it is without the millions of slaves like Mandy who worked its soil. He taught my mother and her brothers to recite these words: "Our white ancestors laid the foundations for this country, but our dark-skinned ancestors built it. They worked the fields, nursed babies, preached sermons, and fought in wars. They played music, owned businesses, cured sickness, and worked on railroads. They taught their children everything important about life in this world. They taught their children about God."

Since my first train ride, I'd had many lessons on dealing with racism, yet I was surprised every time I traveled to the South and saw Ol' Jim shamelessly display himself in the WHITES ONLY or COLORED signs over water fountains and restroom doors. During and after the civil rights movement of the 1960s, those signs came down, slowly, but the hatred and resentment that had hung them there linger.

One day in Boston, shortly before I traveled to Virginia to research my family's history, I pulled out of the parking lot of the hospital where I had worked for nine years and my BMW's blinkers malfunctioned. So I used hand signals. Another car pulled up close, and the driver shouted, "Get your fucking black hand back in your fucking car!" Then he sped off. The coward was gone before I could see his face. I feared worse in the South and, later, saw the behavior of the rude librarian as a warning, but after that, I allowed the exciting discoveries and the welcoming people to dim my vigilance.

On the last day of my trip, looking forward to winding down as I enjoyed a homemade breakfast, I found a seat in the cozy dining room of the B and B. Pat stopped by to say goodbye and thank me for staying with them.

"Lots of cream and sugar, I remember," Bobby, the host, chef, and waiter of the B and B, said with a grin as he poured coffee into my mug. "Don't want it to taste too much like coffee!" he teased.

I leaned back and gazed through the window at the dense grove beyond the well-kept lawn. I thought about Mandy losing her way of life, Coreen losing her child, and Jim losing his home and family just a few miles away. But their survival assured me they must have found a way to believe in themselves, to identify a purpose for their lives. They wanted to leave a legacy and teach values through stories to be passed on to future generations, I decided. For family.

"May we join you?" a man asked.

I looked up. A white couple, probably in their late thirties, stood over me. The man, blue-eyed and blond, was well over six feet tall and wore starched and pressed khaki slacks and a light blue jersey. The woman, a tall, big-boned brunette, also wore casual but crisp clothing.

Though there were empty tables nearby, and I preferred to relax and reflect in solitude, I knew the basis for the B-and-B code of behavior: Every guest is a member of the household.

"Certainly," I replied.

"We're from Baltimore," the man said, pulling out a chair and sitting down. The woman waited several moments, looking at him, then took a seat. I wondered whether she had been waiting for him to stand up and pull out a chair for her or whether she had hoped he would move to another table. I couldn't read her smile.

"I'm from Boston," I responded to the man's statement.

"Really? I grew up in Braintree," he said, opening his napkin and arranging his fork and knife.

"That's just a few exits down the interstate from where I work."

"Been up and down that so-called expressway a thousand times. Had enough. Braintree is an okay town, but it's just a place on the way to somewhere else. I prefer Baltimore. More action, like the Hub, but more real. So what do you do there?" he asked. The woman smiled at him as if she thought his calling Boston by its nickname "the Hub" was clever.

"I'm a pediatrician."

"Is that so? Good for you. I'm a dentist." He patted the woman's knee. "This gal here is retired. She used to be my hygienist; now she's my wife." He guffawed. His wife smiled again. This time I could see she was embarrassed.

Between bites of thick-sliced bacon and sips of fresh-squeezed orange juice, we chatted about what to do and see in the area. I recommended Montpelier but did not mention the purpose of my visit.

During a lull in the small talk, I reached for the pitcher to pour warm maple syrup on my apple waffle. As a stream of clear, brown liquid began to flow toward my plate, the dentist cleared his throat.

"I have to change the subject to something sticky," he said, his face solemn.

The syrup?

His wife looked puzzled.

"Black people and white people can't seem to get along," he said, scanning my face.

"Really?" I braced myself. His wife concentrated on the pats of butter melting on her waffle.

"In Baltimore, a large part of the inner city is black. The mayor is a black man." The dentist smiled, leaning back. "He's liked and respected, so why doesn't he just tell everyone they should put racism behind them and work together?"

"Yes," his wife said, glancing up from her waffle. Her vacant smile penetrated my skin like millions of tiny needles.

"It's complicated," I found myself saying. "Racially based attitudes and behaviors can be subtle and difficult to recognize." I knew I sounded professorial, but I continued, "Racism is systemic — this country wouldn't look the same without it."

"Oh, come on," the dentist said, scooting his chair closer. "Blacks have it made. Kurt Schmoke graduated from Harvard Law School. Now he's mayor of Baltimore. It's a southern town, you know. I voted for him." He grinned. "And look at you. You're a doctor."

I put down my fork.

"Blacks can do anything and be anything they want," he said. "Discrimination is a thing of the past." He tapped the tabletop. "When something doesn't go their way, blacks blame race."

My jaw clenched and began to ache.

"Take the incident with those black Secret Service men in that Denny's Restaurant in DC," he continued. The smell of syrup on his breath made my stomach churn. "They claimed they weren't being served because they were black. I've eaten in Denny's myself; the service is bad. That's all."

"Other customers who came in after them were served first," I said.

"You're wrong. A group of white Secret Service men came in about the same time and happened to be served before the black guys."

"That's not what the newspapers reported. And Denny's ended up losing a class-action suit. The courts fined the company and required it to give sensitivity training to every employee."

The dentist shook his head. "See, this is the problem. This is what I'm saying."

It was my turn to scoot closer. "Regardless of what did or did not happen at Denny's, discrimination is not a thing of the past," I said. "In that same city, the capital, no less, and in many others, black professional men and women wearing business suits and carrying briefcases complain that taxis will pass them by to pick up white passengers a few yards up the street." I could not stop. "Blacks throughout the country experience this kind of treatment every day in restaurants, shops, hotels, at work, on the street, everywhere."

"Calm down. No need to overreact," the dentist said.

His wife looked up from her clean-as-a-whistle plate and dabbed her mouth with a corner of a napkin, trying not to smear what was left of her red lipstick. She smiled, and I thought I saw sympathy in her face. Did she feel bad for me because of her husband's behavior or because of my blackness?

I glared at them. "A few years ago, I was at a medical conference, and a white colleague asked me how it felt to be black. I told him, 'Great. No one should miss it.'" The couple appeared so confused I knew I should feel sorry for them, but my voice rose. "Why would he ask that, and what made him think I could speak for every black person?"

After a brief silence, I continued, my voice unsteady with anger. "This is a bed-and-breakfast. I paid my money, and I'm a guest here, like you. I am not going to be the sounding board on which you can try out your ignorant, arrogant ideas about race relations, and I am certainly not going to validate your notion that racism is only in my head."

Heart racing, temples throbbing, I pushed my chair away from the table, stood up, and headed toward the door. Floor and ceiling pressed toward each other, and rage obscured my vision. The room, the B and B, and all of Virginia had gone silent. As I kept walking, I felt a pair of well-matched, all-American fools staring at my back.

Just outside the dining room, I stopped. I realized that Mandy, Coreen, and Jim—no matter how much self-worth they had or how solid their commitment to family—could not have walked away as I just did—at least, not without consequences. This confrontation with racism would have its emotional costs for me, but I would not be beaten or sold.

I returned to Boston, determined to put the B-and-B incident behind me, but I kept fuming at the dentist's brazen arrogance and the hygienist's smiling complicity. I could not figure out why I let them send me into a tailspin. They were afraid of me and knew they could change neither my history nor my future. The upper hand was mine. In fact, the dentist had offered me the power to wipe away his prejudices, and all I had to do was pat him on the back for voting for a black man.

But well before then, racism had become the flash point in my life, a gnawing entanglement of anger, pain, frustration, and sense of futility that rarely failed to set me off. I had spent decades trying to avoid, confront, resolve, understand, circumscribe, ignore, minimize, and deny the impact of racism on my life, though if any black child could have been protected from racism in America, it would have been someone like me. Like Mary Stone, the squeaky-clean, middle-class teenage daughter in the 1960s TV series *The Donna Reed Show*, I was popular at school and earned good grades. Mary's dad had a successful white-collar job; her mom was a homemaker. My dad was a second-generation physician; my mom was a second-generation teacher as well as a homemaker. But Mary Stone was white. She could believe racism did not exist. I could not. In my well-regarded, racially diverse elementary school in liberal Northern California, white classmates called me "Brownie" or "Fuzzy," and by the time I entered junior high, I knew bigotry would limit my future unless my academic performance was much better than theirs.

The racial prejudice I lived with extended well beyond the

classroom. A few months before I began my senior year of high school, I saw an ad in a local newspaper. A drugstore needed a clerk that summer. I called and scheduled an interview with the owner, Gil. The night before my appointment, I ironed my skirt and blouse, washed my shoelaces, and powdered and buffed my white-suede saddle oxfords, the shoe of choice among teenage girls who wanted to be hip.

When I entered his office, the heavy-jowled man looked me up and down, sprawled out in the chair behind his massive desk, and picked up a lit cigar. After taking a few puffs, he said, "I'm not about to have no colored gal working in my store."

I felt flushed and caught my breath, but then, remembering Gramps saying, "Racism is just another challenge, and challenges make us strong," I looked Gil in the eye and said in a level voice, "Your loss." I turned my back on that man and walked out the door.

Over the years, I learned that slavery and racism had left an indelible imprint, a birthmark, on America. The driver, the dentist and his wife, the drugstore owner, and my white schoolmates decades earlier reminded me that no matter what kind of car I drove, how many academic degrees I earned, how spotless and hip my shoes were, or how multiracial my classroom was, there would always be people in this country who thought I was lesser.

Mandy

*The boat stopped. It swayed and groaned, more quiet now. Over-
head, the trapdoor opened, grinding, creaking, falling back loud and
hard, casting a square of light down on us in the dark. Black outlines of men rushed through the light, down a ladder, shouting. Rods.
Thuds. Whips. Whirls. Screams. A racket of chains. Children crying.
Commotion all around. A whip cut my back, then my shoulders.
Someone shoved me to the ladder. Chain on my arm yanked me up,
chain on my leg yanked me down, a girl's arm caught in it. I dragged
her up. Had to.*

*Outside, above the black hole, chaos. Black men. Black women.
Little black boys and girls, some smaller than me. The bright day-
light made me squeeze my eyes shut. Legs so weak, I crawled. My
knees scraped raw on the splintery wood; they burned in vomit. My
hands slipped in blood. Wiped them on my thighs. Thick and sticky.
I forced my eyes open, and despair poured into my heart. The ocean
lay still and gray and ugly. Not dancing, blue, and beautiful. Not
mine. Not mighty.*

I stumbled down a wood plank, almost fell into the dying ocean. I dropped down beside it and pulled my fingers through my hair. Only one of my pretty red beads was there.

Everything — the sounds, the smells, the sights, the people — was different. White men prodded, beat, whipped us away from the shore, crowded us into pens. Dumped buckets of water on us. Water poured over my head, into my nose, my mouth. I gasped for air, squatted down in our filth. They threw food at us. I cupped my hands, shoved it into my mouth. I gagged.

Beyond the low walls around us, the sun came up, day after day, lifting light and color out of the far edge of the water. And day after day, the sun dropped behind the hills, leaving the sky black.

After a while, most of my sores healed. Then a pale man came and touched me with his dirty hands, smeared oil all over me, made my body shine and reek. My breasts had just started to grow, but he rubbed them a long while. It hurt. I begged him to stop, but he laughed and took his time. Then he slapped my backside and shoved a piece of filthy white cloth at me. It was so small, I could only wrap it around my hips. He flicked my nipples and laughed again.

When he had finished with all the girls and women, the man pushed us into a big, hot room. We were scared. We huddled in a corner. So many white men, talking, laughing, looking at us. I linked arms with the girl next to me, but the man with dirty hands pulled me away, pushed me high up on a box. Head aching, I kept my face down, didn't want to see the men all around me. Smoke from sticks burning in their mouths filled the room, stung my eyes. Some of the men chewed something, spat into pots, spat onto the floor. So many men, so much smoke. Hard to breathe. They came close, walking around me, touching me. I smelled their sweat. They pulled up the cloth, looked under it. I hated that. One climbed up on the box, opened my mouth with a stick, poked the stick through my hair.

A shout. Then more. Faster. Louder. Like drums banging. Every-where. My ears almost bursting.

Then it got quiet. My heart pounded loud and fast.

A man pulled me off the box, put a rope around my chest, and yanked me to a wagon filled with black people. I was glad to get out of the room, but I was afraid.

The bumpy road tossed the wagon up and down. I held on to the sides so as not to fall. My hands pained, but I didn't let go. The sun, hot and angry, cruel and white, clawed my back. I looked over the side, saw the burned red soil. Around us, fields of big leaves rolled like a green ocean.

After a long time, the wagon stopped. The sun was low and or-ange, kneeling and hiding behind a forest. The trees were tall and straight, the branches too far up to reach. So different from the twisted tree that watched over my village and my ocean.

People in patched-up clothes hanging loose on them, rags on the women's heads, gathered around the wagon. Some of them were dark like me, some golden and brown like the wet sand back home. Some had wide noses and full lips like mine; some had thin noses and lips like the men in the smoky room had.

We climbed down. A small boy touched the cloth on my body. A wide-shouldered woman took my hands in hers. A thin woman wrapped her arm around my waist. She wanted to touch the bead in my hair. I let her.

The two women, a girl following behind, took me to a hut. Left me there. It was dark inside, but fading orange skylight came through gaps in the wallboards. I knelt down on a mat, and when my eyes got used to the dark, I saw women and girls standing along the walls, their faces hidden in the shadows.

An old woman carrying a bucket of water came and sat down

on a stool beside me. She lit a candle, and yellow light flickered over her wrinkled face. Reminded me of the elders at home. Pearly rings shone around the black part of her eyes. Her lips were sunken, and when she smiled at me, her mouth opened like an empty cave.

I heard a familiar sound, and when I turned toward it, I saw the old woman's knotty hand sweep a rag through the bucket. The rag splashed, swished, fluttered. Then, like a dead bird, it hung, dripping water through her fingers. She gestured for me to stand up. After I did that, she lowered the sack off my body and wiped cool water over my naked skin. I shivered. She rubbed away the bad-smelling oil and poured a cup of water onto my feet. Caressing two crossed twigs joined with string, the old woman bowed her head and prayed.

In silence, another woman wiped me dry. Another slipped a patched-up, clean cloth over my head and slid my arms through holes in the sides. Another tied a rag around my hair. Motioning for me to sit, the elderly woman set a bowl of mush on the floor next to me. She lifted my hands with hers and placed the crossed twigs onto my upturned palms.

All the women and girls left. I was alone.

I didn't know what that cross lying on my hands was for, but I knew it was important, so I set it on my lap, very careful. I sat there awhile. Then I dipped my fingers into the bowl and let the cornmeal mush settle thick on my tongue. I swallowed. Outside, the orange light was gone. I stayed on the dirt floor, eating in the candlelight. Far away, a woman hummed sad and sweet. Her voice came from nowhere and everywhere, soothing my battered spirit and telling me I was alive. I was alive, and I knew how I would get through whatever was going to happen to me: I would not tell anyone the secret told to me when I entered the world. Never.

8

Beads

My husband and I, both physicians, worked long days. Often, our only time together was a late-night meal. One night in the spring of 1993, Lee and I sat down to dinner at eight thirty. He was quieter than usual and did not seem interested in his broiled lamb chops and sautéed green beans, one of his favorite dinners. When I asked how his day had gone, he let his head fall onto the chair back. Gazing at the ceiling, his voice tinny with fatigue, he said, "The OR was insane; nothing went right. The pharmacy delivered the wrong anesthetic; a surgical clamp snapped open, so the patient's blood pressure tanked; the nurse, a big guy, fainted when a little blood squirted onto his shoes . . . I could go on. And when I got to the ICU, it felt like every patient in there was trying to die." He sat up and pushed his plate away. Staring vaguely at the calendar on the wall, he said, "I'll be fifty this year, but sometimes I feel ninety."

"Maybe we should take some time off," I suggested. "We could hang out on a beach for a week or so. A place with no phone service." Being on call through many nights, in addition to being on our feet eight to ten hours a day, five or six days a week, was wearing us down. I was more than ready for a break from runny noses, ear infections, behavior problems, and distraught parents, not to mention the constant battle with insurance companies.

"I can get time off. Folks know I'm way overdue," Lee said. He leaned forward. "How about Portugal? The Travel section of last Sunday's *Times* said the beaches in the Algarve area have rock formations that look like prehistoric sculptures, and the resorts are secluded and peaceful."

"Sounds perfect." I felt my spirits lift. "I'll get someone to cover the office."

"The resorts probably have telephones," he warned, smiling. "You'll have to leave your pager at home."

"You too."

"We can rest, stuff ourselves, and then maybe go to Lagos." Lee pulled his plate back and lifted his fork. "We might find something for your research there."

"I know that's where the slave trade got started, but I'll have to read up on it before we go." I found myself smiling too. I had wished for more free time to hang out in libraries and bookstores, and I'd been thinking about cutting back my office hours so I could do more research. I was energized by Lee's suggestion. But I was also scared.

Though friends said I was as cool as a cucumber, I was still struggling to put my fury with the dentist and his smiling wife behind me, and I worried that when I set foot on the soil where the transatlantic theft and sale of black people had begun, I would become even more enraged. Yet I suddenly knew that going there was essential. Meeting Carolyn and T.O. in Virginia had helped me see that what I had embarked on was more than a genealogical account of one kinship.

In June, we went to Portugal. Borders of lavender-blossomed jacaranda trees shimmered in the heat as Lee and I meandered down Lisbon's multicolored-tile sidewalks. We dodged traffic and discovered tiny shops jammed with colorful pottery in haphazard arrays on shelves and tables and intricate needlepoint rugs lying across chairs, benches, and floors. We chose a rug with a mauve and Wedgwood-blue floral design. Between us, Lee was the more persistent negotiator, so I stood aside while he bargained with the shop owner and got a good price.

On the recommendation of the hotel concierge, we took a day trip to Sintra. A quaint and rickety wooden train carried us up steep inclines to the storybook town. Castles — some built in the eighth and ninth centuries — crested soaring tree-covered hills and overlooked plunging valleys. Blithely glancing down from one of the tiny balconies on the fifth floor of the ornate Quinta da Regaleira, I felt vertiginous as the bottom of the chasm below seemed to rush toward me. I stepped back.

The following day, we drove along the coast toward the Algarve. Motorcycles, cars, vans, and trucks zoomed around our diminutive rental car while we crept ahead on the narrow and, in some places, barely paved road. Finally, I pointed to the sign, not much bigger than a postcard, it seemed, for our resort.

For the next few days, Lee and I lounged among startling orange, red, and gold rock formations jutting out of glistening, pristine shores. I could feel my muscles relax in the warmth of the sun. For most people, beach food is salads and sandwiches, but once we'd tasted the Portuguese cabbage and potato soup in the dining room on our first night at the resort, we could not get enough of it. The waiters thought we were strange to want hot soup in hot weather, but they brought soup to our seaside lounge chairs every day.

After a few days, Lee and I decided to leave our beach refuge and let our trip take a more serious turn. Five centuries earlier, the citizens of Lagos had seen the first ship sail toward the African continent, beginning the devastation of millions of African lives.

I knew from my research that the governor of Lagos, Prince Henry the Navigator, was a patron of exploration. His aim was to spread Catholicism throughout the world, but another, less virtuous, motive was to find a sea route to the ivory and gold of sub-Saharan Africa.

In 1441, Prince Henry sent two ships to Africa. They landed on an island off the coast of current-day Mauritania. The explorers came upon a market crowded with black Muslims and captured twelve of them. Among the captives were two Moors of noble birth who offered "ten blacks, male and female," in their place. The twenty Africans were taken to Portugal, shown to Prince Henry, then auctioned off in Lisbon.

In a series of voyages, the prince's men explored the African coast and discovered large urban complexes governed by dynasties as well organized and technologically advanced as those in Lisbon and other European cities. A vigorous trade relationship quickly burgeoned between African and Portuguese merchants. In the trade's early years, European goods, including brass, pottery, and linen, were exchanged for African treasures, including ivory, silver, and, especially, gold. But within one generation, Africa's stolen people became the greatest treasure of all. The slave trade was so lucrative that even the allure of gold was all but forgotten.

At first, Portuguese ships transported captives to Europe to be Christianized, but this practice was short-lived because it slowed the flourishing slave trade. Besides, the Portuguese reasoned, if their captives were not Christians, they were not human beings, so they didn't need to feel guilty about selling them.

No other event in the history of the world would come close to causing the destruction that this dispersal of men, women, and children wrought. And nothing would rival the wealth and power that the trafficking of human flesh reaped for the merchants and slave owners.

Lee and I prepared for our visit to Lagos with a history book, a local map, and a travel guide. The paragraphs of the history book describing the arrival of the first African slaves in Lagos were highlighted yellow. A red line on the map marked our route from the resort. Sticky notes along the pages of the travel guide flagged descriptions of churches, the customs building, town hall, and other tourist sites. The single blue tag marked what was, for us, the most important destination: the stockades. The guidebook, which I had purchased in the remainders section of a bookstore, briefly described their massive iron posts, but I could not visualize them. I had to be there to see and touch the stakes to which thousands of slaves had been chained.

As we drove to Lagos, about ten miles west of our resort, I noticed Lee holding the steering wheel tighter and tighter, and, squinting at the dusty road unfolding before him, he seemed unaware of the passing scenery. In the heat, the map I had laid across my lap felt like a damp blanket, and I could feel tension building in my neck. We spoke in two- or three-word phrases, able to communicate only as driver and navigator.

The trip felt long, and there was little evidence that the landscape had changed since the first slaves set foot on Portuguese soil. To the north, the trees were thin and parched, the hills behind them barren. To the south, roadside boulders stood against the backdrop of the flat, still ocean.

"I think this is the turnoff," I said at last, and Lee steered the car onto a narrow road. Time seemed to speed up. Suddenly we were in Lagos facing a row of larger-than-life bronze statues of Henry the Navigator and various other Portuguese aristocrats, all in formal regalia. Bronze conquerors armed with shields and swords stood among the nobles as if protecting the tree-lined harbor, the historic locus of Portugal's wealth and power. We parked at the feet of a massive figure with a commanding posture and an expressionless face. The tightness in my neck spread to my shoulders.

The town was picturesque, tranquil, but as we walked along narrow cobblestoned streets, the old city — with its remnants of defense walls, its ancient buildings with crumbling cornices and rotting doors — seemed to close in on us. In the Praça da República, a tidy scene of grass, trees, flowers, and walkways, I tried to take a deep breath, but now my chest felt tight too.

Lee and I circled and crisscrossed the plaza looking for the iron posts of the stockades. The more we searched, the more I sensed the haunting presence of slaves — skeletal warriors slumped against the stockade, shackled mothers who would never see their sons and daughters again, and children curled up on the dirt, sobbing. It felt as though their ghosts followed us.

For almost an hour, we searched, referring to our travel book for the location of the stockades. At one corner of the plaza we found the stone Customs House, the site of the African slave market, but no trace of the iron posts. Finally, we stopped in the center of the plaza. An enormous red wagon with a garish yellow awning faced us. Bottles of soft drinks were lined up on the counter, and colorful bags of potato chips and pretzels dangled from wires. We estimated the distance from the harbor; Lee and I looked at each other, then checked our guidebook again.

We stared at the concession stand and the vendor wiping the counter, arranging bottles of soda by flavor, and clipping more bags of junk food to the wires. This was where the stockades had been.

The sunny town square suddenly felt sinister in its indifference, everyone around us complicit.

A young man in a colorful T-shirt and snug-fitting khakis sauntered by eating a thick roast beef sandwich, allowing bits of crust to fall onto the sidewalk.

In front of an Old World café near the concession stand, a slender gray-haired man in a stylish linen suit lounged back in a metal chair, smoke rising from his cigarette.

Across the plaza, a woman wearing a bright flower-print dress smiled as she watched her two sons run and jump, squealing with joy, through the four arches of the Customs House, where slaves had been bartered and sold.

I felt stupid and naive. I had attributed to the people living in Lagos an intimate understanding of the world-changing atrocities that had occurred in the city they called home. But the inhabitants seemed dismissive or oblivious.

"Damn it!" I shouted.

"Let's get out of here," Lee said, folding the map and shoving it inside the guidebook. Perhaps the Portuguese did not believe their ancestors' participation in the slave trade important to remember. Or perhaps the truth was too painful to face. The residents of Lagos might not have realized that history could not be tidied up so easily.

We rushed to the car, passing the bronze nobles and soldiers on the way. I looked up at Prince Henry and his cohorts, men who had caused human suffering on a scale so massive, it was impossible to measure. These were the men Lagos memorialized as heroes. Not a single plaque acknowledged the slaves.

Lee and I could not speak for a long time, and when we reached the resort, it felt as though the colorful gardens, groomed hedges, and soft music drifting out from the lobby betrayed what had happened to thousands of slaves just ten miles away. Lee retreated to our room. I went for a walk, hoping that fresh air might relieve the ache in my neck and shoulders.

As I passed the large windows on the façade of the hotel, my own reflection startled me. I saw a middle-aged African-American woman wearing a white short-sleeved blouse neatly tucked into sharply creased navy-blue slacks. Her short, curly hair was stylish, her posture and movements casual. History had been kind to her.

I headed for the deserted beach. In the peacefulness of the approaching evening, the beauty and luxury of my surroundings, and the frivolity of the beach chairs and multicolored umbrellas, I felt guilty. The family directive had served me well. I enjoyed the social mobility and financial security that being a doctor afforded me, and my family was so very proud of me. I waltzed through life like the presidential descendant that I was, seldom giving a thought to what my enslaved ancestors and their children and grandchildren had done for me. I had taken freedom for granted.

I, too, was complicit in ignoring the suffering of my African fore-bears.

Standing at water's edge until nearly ten o'clock, I watched the sun set and the sky fold into layers of gold and fuchsia. On that clear evening, the ocean darkened, the edge of the horizon so sharp it seemed within reach. An engine-powered ship crept west-ward through the water, and in its wake, the sound of erupting waves shattered the stillness of dusk.

I watched the modern ship and envisioned in its place a slave ship, its sails ready to sweep the weather-beaten vessel toward the New World. The ancient boat hovered just beyond the beach, and massive stockades stretched from the nearby fifteenth-century Portuguese village of my imagination to the ocean, its horizon swallowed by dense fog. From among the stockades' iron posts, the ghosts of thousands of stolen men, women, and children, includ-ing toddlers, emerged. The mist ebbed and flowed around them, but I could see their eyes, vacant, bewildered, tear-filled, enraged, hopeless, terrified, or defiant. I extended my hands toward them.

Standing there, as I began to work through some of the shock and pain of that day, I realized I had gained an understanding of why the *griots* of West Africa are so necessary. Through oral history and family keepsakes, our African ancestors are not mere phantoms. They are alive in the stories told from generation to generation, in the letters and documents kept in family Bibles, in photographs inside old cardboard boxes, and in slide-show images on makeshift screens. They are alive in the ambitions, perceptions, beliefs, and values of their descendants, the people who are the evidence of what happened to the stolen Africans who survived. I would become the *griotte* for those families as well.

Alone in the growing darkness, I recalled hearing a report that sea-eroded beads, thought to have been worn by Africans whose dead bodies were thrown off ships en route to the New World, had washed up on the shores of West Africa. I envisioned a path of red beads undulating with the currents on the ocean floor and tracing the Middle Passage from West Africa to America, the path Mandy had followed from Ghana to Virginia. The beads were like the red line on the map that had led Lee and me to the concession stand, and they reminded me of the blood Africans spilled in their homeland, in the sea, and in the New World. Beads were among what little the captives could hold on to when they left behind everything that had filled their lives: music, dances, customs, ceremonies, friends, families, and lovers. The glow and heat of the sun over Africa. The caress of their own soil on their hands and feet. The anticipation of a familiar tomorrow. The sea-worn beads begged to have their stories told.

Soon after I returned from Portugal, I went to the library to learn what had become of the stockades. The updated edition of our guidebook made no mention of the stockades and their iron posts. I could not find a morsel of information. The erasure was complete.

9

The Castle

The blue of the African sky was the most perfect color God ever made. The air was so clear and the sunlight danced so brightly off the ocean, I felt as though I were standing inside a crystal. Where the ocean swept over beach, the soft golden sand became firm and copper-colored. I ventured into the water, and it curled over my feet, soothing away the heat of that African day. I stepped back onto the dry sand and turned to watch the water gather up my wet footprints, taking them into the ocean's depths, disappearing westward — as others had gone. In the distance, one cloud floated from the west — as I had come. It was 1995, two years after my first trip to Virginia, and I was in Ghana with a group from my church. My aim was to trace Mandy's footsteps from Africa to America.

I stood on the shore, my shoulders hot under the powerful rays of the sun, and looked up at Elmina Castle. It stretched along the shore and rose out of the rocks like a white crown, its series of spires and tiers of walls and balconies embellished with arches and studded with cannons. Long canoes painted with bright colors in bold designs crowded into the small harbor alive with voices, music, and fish in nets flopping against the boats. Beautiful in spite of its history: Elmina, Ghana.

From the fifth to the eleventh centuries, Islamic merchants referred to the empire of Ghana as the "Land of Gold." The name Elmina comes from the Portuguese words *a mina de ouro,* "the gold mine." It was probably during the Netherlands' control of the area, 1637 to 1872, that the labyrinthine complex of cave-like rooms took on the appellation *castle*, a derivative of the Latin word *castellum*, meaning "fortress." Under British rule, from 1872 to 1957, the West African colony was known as the Gold Coast; in 1957, the region gained its independence and took back its ancient name, Ghana.

When Portuguese explorers reached Ghana in 1471, they were greeted by natives adorned with gold jewelry. These upper-class Africans owned slaves. Slavery was an accepted social institution throughout Africa, but the slaves were usually prisoners of war, debtors, or convicts, and their enslavement was neither perpetual nor heritable.

Initially, Portuguese bounty hunters traveled relatively short distances inland to abduct or purchase natives to sell in Elmina, but when the demand for slaves expanded, the Portuguese, with the aid of African henchmen, began to kidnap natives throughout West Africa for transport to Europe and, soon, to the New World.

When the Portuguese proposed building a permanent base in Elmina, the local chief, Caramança, tried to discourage the plan. Nonetheless, in 1482, Portuguese soldiers hoisted the flag of Portugal and banners of the Catholic Church. Pope Nicholas V had issued Romanus Pontifex in 1454, a decree that granted Portugal's King Afonso V the right "to invade, search out, capture, vanquish, and subdue all ... enemies of Christ ... and all ... goods whatsoever held and possessed by them and to reduce their persons to perpetual slavery." Under this authority, Portuguese soldiers claimed African territories and forced captives to swear allegiance to Portugal, renounce Islam and traditional African religions, and convert to Christianity or be made slaves. Under duress, Caramança accepted the sovereignty of the Christian God. The edifice that became the hub of Portugal's slave trade was originally called São Jorge da Mina, named for Saint George, the patron saint of Portugal — and the saint of those who suffer.

Some newly captured slaves were made to walk hundreds of miles to Elmina; others were brought in boats along rivers or up the coast. Within a decade, at least ten thousand African men, women, and children, many branded with a cross, became prisoners in the fortress. Some young people came willingly; their parents thought they were putting their offspring in the hands of benefactors who would take the children to lands where they could have better lives.

In 1637, the Netherlands, with tactical assistance from Africans eager to rid themselves of the Portuguese, captured Elmina. Little did the Africans know that the Dutch would be far more oppressive than the Portuguese and would take the transatlantic slave trade to its zenith.

When I walked through the entry into the central courtyard, the buildings, briefly, seemed majestic. Graceful balconies offered sweeping views of the ocean, and shady porticos granted relief from the sun. The large square was open to the sky, but I felt closed in. Everywhere I looked, I saw crumbling plaster walls, interrupted high above me by rows of windows, black and opaque from where I stood on the cobblestones below. Along the walls, dozens of cannons pointing in all directions stood at the ready, stacks of cannonballs nearby.

In the central courtyard, I joined a group of visitors from distant parts of the world. I was surprised when our guide, in his melodic, punctuated baritone, informed us that the building at the center, with its square façade and square windows, was a church, the first Catholic church in sub-Saharan Africa. When the Protestant Dutch took over, they refused to worship in a Catholic church and used the building to buy and sell goods and slaves. These new occupants built a church of their own, a Protestant Dutch Reformed church, on a higher level of the castle.

Although this Catholic church had lost its function as the center of religious life at the fort more than three hundred and fifty years before my visit, its presence felt starkly at odds with Elmina's brutal role in the slave trade. In the system of dark tunnels and dungeons below the church, thousands upon thousands of Africans had been chained, beaten, starved, and herded like cattle before they were loaded onto ships that would carry them to enslavement in the New World — if they survived the Middle Passage.

Near the church there was a tiny cell that had been set aside for slaves who had been condemned to death, most often men who resisted captivity. There were only a handful of us on the tour, but when the heavy door slammed shut behind us, enclosing us in utter darkness, I felt crushed among hundreds of bodies. The sound of each person's breath echoed hollow in the tight room.

In the dark, our tour guide revealed what had happened here. Ten or twelve men at a time, he said almost in a whisper, were whipped, bludgeoned, shackled, and then dragged into this cell. The door was locked. No food, no water. Every few days, armed guards would look through a small opening to check the condition of the prisoners. Still no food, no water. The guards came back until, finally every prisoner was dead. Our docent's soft, rhythmic voice was like a hymn of mourning wafting from the church across the plaza.

On the other side of the courtyard there had been a holding area for female captives. The guide pointed up to a balcony and told us of a "tradition" at Elmina: Because there were few European prostitutes in Africa, when a white man wanted women, he chose one of the female captives. The Dutch governor of Elmina had first choice. He would order his soldiers to bring a group of prospects to the area of the courtyard below his living quarters. These were among the few times the women could leave the cells, and at first, they might have welcomed a breath of clean air and a view of the sky, a reprieve from the dungeons' stench and darkness.

The governor would step onto his balcony, survey the women below, and point to the one he wanted. With everyone looking on, guards dunked the chosen woman, again and again, in the rectangular cistern of water at the center of the courtyard until she was clean enough for the governor to touch. Then the guards pushed and prodded her up a ladder and through a trapdoor that opened directly into the governor's bedroom above. The trapdoor closed and locked.

When the governor had satisfied his needs, the woman was isolated in a cell for a month. If she had become pregnant, she was taken back to her village or one nearby. Perhaps she believed that this time, she truly had earned a reprieve. The woman might not have known that a few years later, the guards would come to the village to take her and her child into captivity.

═══

I stood in the courtyard, imagining myself naked and huddled among a score of African women. We were all thin and weak following days of hunger and thirst. Our bodies were tall or petite, narrow- or broad-hipped. Our breasts were small, full, erect, or flaccid. Our faces were round or slender, our foreheads low or high. Our cheeks were smooth or cut with tribal markings. Some of us cried; others frowned; some cowered and trembled. Some of us tried to hide our nakedness; others knew trying to hide was useless. Some had short-cropped hair; others had rows of braids; some had beads laced into intricate hair designs. Our beads were yellow or blue or green or orange or, like Mandy's, they were red.

The docent led me and the other tourists through a doorway that opened into a cave-like room. The air was sour, musty, and moldy. As my eyes adjusted to the dim light, I realized bats, perhaps a hundred, hung from the arched ceiling. I must have looked alarmed because the guide assured me they were harmless. He called the area around us "the room of deep sorrow." He extended his arms toward the black openings on opposite sides of where we stood and told us these were the entrances to tunnels. One led to the men's dungeons, the other to the women's. In this room, men and women said goodbye. Here, husbands and wives, fathers and daughters, and mothers and sons were separated. Most never saw one another again. The anguish, the fear, the sense of loss, and the sorrow all lodged in my chest.

Through somber light and looming shadows, we followed one of the damp tunnels to a narrow stairwell that ended in the dungeons below. Along a dimly lit corridor, a succession of arches opened into the female quarters. Originally storerooms, the guide explained, each windowless space could hold one hundred and fifty to three hundred women. Often, they had to stand for days or, sometimes, weeks at a time, skin to skin, mired in feces, urine, vomit, pus, tears, and menstrual blood. Many would die here.

When the time came to board the ship, the women who had survived were led through an interconnecting series of rooms to a final chamber that offered the only sunlight many had seen for weeks. Blinded by the light, the women entered a narrow, gated egress that led to the beach. The intensely blue sky draped over their heads, and maybe the women glimpsed loved ones as their eyes adjusted to the brightness.

They had stumbled through, crawled through, or fallen through the Gate of No Return. And there, they could feel something familiar under their feet — African sand. Only moments later, the sensation was lost forever. The ocean wiped away their footprints.

10

The Museum

Y ou have to come down here. There's a slave ship in Baltimore. Don't do another thing until you see it," my friend
Bonnie insisted over the phone.

"A slave ship? Really?" I asked.

"I could say it's just a replica, but it feels real — too real, actually.
I could try to describe it, but you need to see for yourself. The
ship's in a museum. We can go together. I want to be with you."

After Elmina, I wasn't sure I was up to visiting even a replica of
a slave ship, but I knew I had to go. I was glad I would be going
with Bonnie. When a mutual friend introduced us several years
earlier, Bonnie and I felt as if we had grown up together, though we
had different interests. She was a nurse who loved poetry, painting,
gourmet cooking, interior decorating, and flower arranging; I was
a pediatrician who enjoyed dancing, gardening, writing, reading,
and collecting arts and crafts. Soon after that first meeting, we
started calling each other "sister."

Two weeks after's Bonnie's invitation, on a Saturday afternoon,

I arrived in Maryland. As she drove me from the airport to her home in Bowie, snow began to fall thick onto the roads. It swept across the windows as the car crept along, its wipers whipping back and forth and its tires slipping. Bonnie turned up the radio. The announcer stated that the airport would soon shut down and that many highways were already closed. We followed a snowplow down the side roads and finally reached her house.

The following morning revealed a white-blanketed town. I assumed we were snowed in. "No museum for us this trip," I said, knowing I had patients to take care of the following Monday.

Bonnie, her hands on her hips, said, "Forget the weather. Don't be a chicken."

Don't be a chicken was not exactly eloquent, but I gave in. While Bonnie and I bundled up in coats, scarves, gloves, and boots, her husband, Curtis, warmed up the car. As we started to pull away, Curtis asked through the open window, "Bonnie, are you sure the museum isn't closed?"

"It better not be," she responded, backing out of the driveway.

"Why don't you call first?" Curtis shouted. A concerned look on his face, he waved goodbye.

After crunching and sliding over some twenty-seven miles of icy roads and taking several detours, Bonnie and I reached the museum. It was open. Located in a renovated firehouse on Baltimore's east side, the National Great Blacks in Wax Museum is a source of pride for the surrounding inner-city neighborhood. The exhibits depict black people from the time of ancient Africa through the modern civil rights movement and up to the present, showcasing the significant contributions made by men and women of African heritage throughout the world. Each wax figure has been researched to replicate facial expressions, posture, dress, jewelry, and various accoutrements. The museum's slogan is "Taking you through the pages of time."

Bonnie guided me toward the slave ship. Plaques on the anteroom walls explained what happened before the voyage: Merchant representatives and members of the crew stripped the prisoners bare for inspection. Mouths were pried open, limbs were manipulated, backs were jabbed, and genitals were palpated. Red-hot brands were pressed to the chests, breasts, legs, arms, backs, or hands of the captives who passed scrutiny. The letters and numbers indicated the individuals who had delivered the slaves and the ships that would carry them away. A subsequent branding designated the trading company that had purchased the cargo. Later, these markings were used to identify slaves in records of sales, legal documents, and official proceedings. Names were irrelevant.

Prisoners with any infirmity or deformity were decapitated. In the wax museum, a disembodied head hung from a pole.

Though disconcerted, I moved on, Bonnie close behind. As we crossed a plank and descended a stairway, a recorded voice announced, "Here's our next group. Looks like a healthy bunch." We were to imagine ourselves as prisoners like Mandy or the other tens of millions of Africans herded onto ships that made some forty thousand transatlantic voyages.

My voyage began. A cacophony of water splashing against the ship, the clanging bell, creaking hull, scraping chains, screams, coughs, retching, sobs, and deep, agonal moans echoed through the stuffy room. I knew the sounds were taped, but the walls seemed to press in on me.

Bonnie guided me to the first exhibit. I was not prepared. A fully clad white man, his fist raised, his face contorted with malice and lust, loomed over a naked black woman. The ruffle of his shirt cuff dangled above her face. The captive's eyes and mouth were open wide with terror. Her outstretched hands pleaded for mercy.

The scene before me had happened thousands of times aboard slave ships, but in this frozen moment, the man did not thrust himself into the woman's body. He did not rip open her rectum or gag her with semen. He did not bloody her nose and mouth with his fists or pierce her breasts with his teeth. He did not slap her into unconsciousness. He simply hovered. Yet in my mind, I saw each brutal act; I heard the woman's screams, the thuds to her pelvis and buttocks; I smelled the nauseating, incongruously sweet scent of blood. I felt as if I, too, were suspended in time, destined to stand less than three feet from this rape for eternity. Or, two centuries earlier, I might have been that woman. I could not turn away.

Bonnie wrapped her arms around my shoulders. I had not realized I was shaking. After several moments, we moved on to the next scene, a row of boys, chained together, sitting in a narrow, isolated alcove. Their mouths hung open. Tears pooled in their eyes. Shining drops lingered on their cheeks.

In the eighteenth century, approximately 25 percent of captives on slave ships were children, but by the nineteenth century, 40 percent were children. Children were easier to capture, control, fit into tight spaces, and overwhelm, and when they died, they were easier to discard.

Slave merchants and ship's officers, among the prime beneficiaries of the slave trade, had their pick of the women, girls, and boys. For the pleasure and convenience of the captain and other men of rank on board, the women were held in quarters immediately below those of the ship's officers. The crew members — many of whom had been recruited from poorhouses, jails, or insane asylums — also ravaged captives, especially boys. *So this,* I thought, *was why the boys were sequestered away. These terrified children; they cried for their homeland, for their mothers. Our boys, not yet men.*

In a heavily guarded section of the ship, African men lay shackled together, ankle to ankle, wrist to wrist. Their separation was intended to keep them under control and hide from them what was happening to the women and children elsewhere on the boat. The crew lived in constant fear of being overpowered by enraged African men.

Slave insurrections, motivated in part by the prisoners' belief that their captors were cannibals, took place in one out of every eight to ten voyages. Most failed. Many of the rebellions occurred shortly before or after the ship set sail. Those on the open sea were the most dramatic.

One such uprising occurred in 1532 on a Portuguese ship named, of all things, *Misericordia* ("Mercy"). The ship was en route from São Tomé to Elmina when the 109 slaves aboard rebelled and killed all but three members of the crew. The three survivors escaped and rowed a small boat to Elmina. The ship and the Africans, however, were lost at sea.

Bonnie and I continued through the museum, taking in representations of slave auctions, the financially crippling and socially demeaning post–Civil War sharecropping system, and a black woman drinking from a rust-stained, cracked water fountain during the Jim Crow era. We saw true-to-life wax figures of African-

American heroes of the nineteenth and twentieth centuries, including Sojourner Truth, Booker T. Washington, W.E.B. Du Bois, and Martin Luther King Jr. I looked at each exhibit and read each plaque, but I could not keep my mind off that black woman's rape.

After more than two hours in the museum, Bonnie and I left and began the ride back to her home. Since entering the slave ship, we had not spoken.

Finally, she asked, "You okay?"

"I don't know."

The winter clouds had vanished, but the late-afternoon air remained so cold, the car windows fogged over. Bonnie turned up the defroster, but still, we could not see out. She pulled to the side of the road and waited for the windows to clear. We shivered, our breaths cones of white mist.

The screams and moans from the slave ship persisted in my mind. They drowned out the whirring of the car fan and the droning of the engine. I felt entombed, trapped in a ship's hold. I wondered if Bonnie did too.

On average, fifteen to twenty of every one hundred Africans died during the Middle Passage, but the mortality rate was sometimes as high as 50 percent. The captives, naked and dehumanized, suffered unconscionable cruelty. They were chained, shackled, beaten, starved, mutilated, and raped. Some were tortured with thumbscrews, iron collars, or red-hot metal prods.

Though some crews tried to minimize the rate of death, disability, and injury among the human cargo, there was always some "wastage" that was tossed into the ocean as the ship sailed along. No matter how "humane" the crew, many Africans died before the ship reached its destination. But millions of stolen people—

those, like Mandy, who called forth profound physical and mental fortitude — managed to survive the voyage, only to begin a life of bondage.

Would I have made it to the New World? In the museum, I'd imagined myself not as an observer of the rape, but its victim. I'd envisioned lying among up to a thousand fellow hostages, many dying of dehydration, measles, smallpox, dysentery, cholera, malaria, or pneumonia. We lay stacked between tiers of wooden platforms only two to four feet high. But I could not imagine being so desperate for air that I would strangle or bludgeon fellow prisoners to death. That level of anguish and fear was far beyond any emotion I knew.

Some prisoners hanged themselves, refused to eat and drink, or jumped overboard. And when no other means of death was available, some captives ripped out their tongues and choked themselves. Some went mad. Would I have been so desolate? Would I have risked my life to help the children, the most fragile of the captives?

From time to time during the transatlantic voyage, the crew forced the survivors, the "lucky" ones, up to the deck to dance for the amusement of the captain and other officers. I closed my eyes and had visions of haggard captives dancing to the hesitant percussion of an African drum while the shrill voices of a chorus of fellow prisoners sang in myriad languages and dialects. I heard cat-o'-nine-tails snap against the backs and legs of dancers who were reluctant, sluggish, ill, or simply not agile in order to keep their steps lively, their voices loud. Had Mandy danced and sung? Would I have been among the performers?

During my visit to Virginia, Ann Miller told me that the stench of approaching slave ships reached harbors miles away. At the end of the voyage, which could last anywhere from one to nine months, the ship was a scene of living skeletons, stumbling and crawling.

Near the shore, white men took turns lashing the survivors with bullwhips to make sure every prisoner was beaten into submission. The sign over the exhibit of a captive being whipped read AT LEAST THIRTY PERCENT DIED DURING THE "BREAKING" PERIOD. In terms of waste, the slave trade was remarkably inefficient. But Mandy survived the transatlantic voyage. She endured the breaking period. She withstood enslavement.

I felt only a small portion of the fear, pain, and loss my ancestors must have suffered as the ship carried them away from everything they knew and everyone they loved. There had been moments when I was in the vessel's hull with my forebears, lying there shrouded in darkness and overwhelmed by the misery, terror, and death around me. The feeling was devastating. I felt I had finally absorbed the meaning of the part of my family credo that says "You come from African slaves."

11

"Visiting"

I never heard my mother use the word *sex*. For her generation, people born in the early 1900s, sex outside of marriage was a sin. In the 1950s and '60s, when I was growing up, there were topics no one talked about in public. Television, film, newspapers, and radio followed a code of morality too. It was a world without social media and the internet; unlike today, images and language steeped in sex were not ubiquitous, and privacy was not obsolete.

For my mother and other black women who had escaped the Jim Crow South, and for those who had not, there was also the multigenerational memory of the predatory white male. African-American mothers faced the quandary of how to maintain modesty and decency while warning their daughters about unwanted sex.

Mom often told me stories about our ancestors while she sat at her sewing machine. I recall one I heard when I was eleven.

"What was Coreen's father's name?" I asked as I stood waiting for my next round of fittings.

"He was James Madison Sr. His son, the one who became president, was James Madison Jr." I can still see Mom, hunched over the machine, seeming a little nervous as she continued to sew.

"How did Mandy and the president's father meet each other?"

Mom explained that Madison Sr. owned colored people. They were his property — his slaves. They had to follow his every order, take care of his every need, and fulfill his every desire. Mandy was one of Madison's slaves. He was her master. The slaves called him Massa.

"Whenever he felt like it," Mom said, "Massa could walk into any cabin where slaves lived and visit whichever woman he wanted."

"Visit?" I asked.

"Yes. Massa went from cabin to cabin. That's the way it was back then."

"Like gentlemen visited you before you met Daddy?" I said.

"What I'm talking about is very different," she said, her eyes on her sewing, her foot pressing harder on the pedal, the needle a flashing blur.

"What's different?" I asked, struggling to understand what Mom was trying to say.

"What happened to Mandy."

"What happened to her?" I persisted.

"Mandy attracted the master's attention because she could pick cotton fast."

"Did Mandy get a reward for being a good worker?"

"No. Massa punished her."

"Why?"

Mom looked up at me. "She was his slave."

"How did he punish her?"

Mom hesitated and looked down again. "I'll tell you later. Not now. You're too young."

"I'm almost twelve," I complained.

"I know." She stopped sewing. "When I was your age, I was curious, just like you."

Gramps was a vivid storyteller, Mom said, but whenever she asked him about Mandy and the massa, he would change the subject. One day, Mom decided to ask her aunt Laura, Gramps's oldest, and rather grumpy, sister. They were alone in the kitchen. Aunt Laura was facing the stove, frying chicken in a big iron skillet.

"Aunt Laura," my mother said, "what did President Madison's father do to Mandy?"

Aunt Laura whirled around, the chicken sizzling, the grease popping behind her. She was a big woman; now she looked even bigger. "Ruby," she said, glaring and waving a long fork, "don't bother me with that kind of question. I don't remember none of that old stuff. Besides, you're too young. Forget about it."

As I stood beside her sewing machine, Mom looked me in the eye and said, "My aunt was right. Eleven is too young."

For help with teaching my brother and me about sex and other important day-to-day concerns, Mom turned to Jack and Jill of America, a national organization founded in 1938 to inform, encourage, and inspire black youths. Our parents enrolled the two of us in the Oakland chapter. Meetings took place in members' homes, safe places to approach uncomfortable issues. Though Jack and Jillers had serious discussions, we also had fun. We bowled, played tennis, and went horseback riding. The moms took us to lectures, museums, operas, ballets, and concerts. The dads showed up at the picnics.

The first time I heard the word *sex* was at a meeting in 1956. I was thirteen. The early-teens group had gathered in the Waltons' house, an expanse of stuffy rooms overfilled with ornate furniture encased in thick plastic slipcovers. My mom's close friend Edwina Walton, a tan-colored woman with tightly curled graying hair, wore an ill-fitting mauve dress, black flats, and a strand of pearls. Her favorite topics were "proper behavior" and "good morals." It was often difficult to figure out what she was trying to tell us. She was an articulate, haughty woman, but that day, standing in the center of her harshly lit living room, her voice was shaky and halting.

"Who knows how babies are made?" she asked out of the blue. My mother had not told me what the topic of the meeting would be.

We all looked at each other. Leon, a rambunctious, athletic boy wearing a blue blazer, his shirttail hanging out, threw up his hand.

"You go all the way," he announced.

Surprised, Mrs. Walton asked, "What does that mean?"

"It means you do it." Leon grinned.

There were a few snickers.

She persisted. "Who does what?"

"A boy and a girl have sex," he answered.

Mrs. Walton turned red, stiffened, and stepped toward him. "That's an ugly thought, young man." Leon slouched in his chair. "A man and a woman get married, and then what they do together is called 'making love.'" She continued, placing her hands over her heart. "And it's a beautiful union of their hearts and souls."

That's weird, I thought, the backs of my bare arms sticking to the hard plastic of the sofa.

Some of my fellow Jack and Jillers looked amused, others confused, like me. Not one of us said a word. I didn't believe Leon knew what he was talking about, so I thought Mrs. Walton was going to answer the question herself. But she turned away from Leon, clapped her hands, and said to the group, "Mildred Thompson will hold the next meeting, and you will have a debating contest about the pros and cons of the freedom of speech."

Linda, who was a little older than the rest of us, said, "We should choose teams by sex." She giggled, and a few others joined her.

I held my breath as Mrs. Walton twisted her necklace and glared at Linda. "The children on my right, team A, will discuss the pros; those on my left, team B, will discuss the cons," Mrs. Walton said. "You may use any research materials you like — library books, encyclopedias, magazines, newspapers, and so forth. Each group will choose a captain, and he or she will come up with an outline of the key points of the team's argument. That's what you will talk about next time."

The meeting was over, and the only thing we had learned about sex was that grown-ups didn't know how to tell us about it. As we gathered our things, Mrs. Walton looked relieved.

Over the years, the Jack and Jillers met at different homes and discussed many topics, but never again did we talk about sex. From time to time, my mother would try to talk to my brother and me about it, but her message was elusive and often odd. I remember her saying, "After Mommy and Daddy got married, we slept together, then, nine months later, Bettye, you popped out, and seven years after that, Biff, you popped out too."

For my fifteenth birthday, Mom and I redecorated my room. I chose pink paint for the walls, and she designed the curtains, bed covering, and skirt around the vanity, all in white organdy with wide ruffles along the edges. As a finishing touch, Mom bought a white rug with pink roses.

One evening, Mom tapped on my door.

I felt grown up sitting at the kidney-shaped dressing table with its three-way mirror and reflecting tabletop, arranging bottles of cologne and trying out different shades of lipstick. "Come in," I said.

Mom opened the door as I was applying Persian Melon to my lips. She wore a pale blue shirtwaist dress and a floral-print apron.

"That lipstick is perfect for a Sunday-afternoon lady," she said, standing beside the dressing table. "It's not gaudy like Saturday-night gals wear. Those Jezebels smear on a ton of makeup, squeeze into low-cut, skintight dresses, then go to nightclubs and try to attract men."

I waited for Mom to tell me what happened next.

"Dinner's about ready," Mom said, turning toward the door. She paused. "I was a Sunday-afternoon lady. When a gentleman came to visit me, he brought flowers, and we sat in the living room with my parents. If he did not greet my parents properly, he was not allowed to come again."

Mom and Dad expected my brother and me to make them proud and advance the gains that they, and the generations before them, had made. Dad was a founding board member of the First AME Church in Oakland. Every Sunday, I attended service wearing a freshly ironed dress, a hat, white gloves, polished shoes, and sheer stockings. Once home, I could remove the hat and gloves, but I stayed dressed up all day. And I always sat with my ankles crossed.

As Mom sewed and recounted stories from our family saga, I, without realizing it, memorized the many small details. Mom's narration was usually animated, but when she talked about Mandy becoming pregnant, she didn't look up from her sewing; her voice was flat, her explanation stunted. She said, "Mandy was not a Jezebel, but she wasn't married when the massa made her have a baby." I knew something was missing. The stories *not* told were as much a part of my family history as the stories told again and again.

When I was almost eighteen, Mom helped me prepare for the senior prom. After I'd gone from shop to shop and tried on dozens of dresses, I discovered a pale blue organdy gown in my mom's closet. I pulled it out and admired the flower appliqués on the bodice and above the hem.

"Why don't you wear this anymore?" I asked.

"It's getting a bit too small."

"Can I wear it to the prom?"

"Of course, but I'll have to take it in to make it fit that figure of yours."

I stood beside her sewing machine, pleased that I was going to wear such a pretty dress. I'd noticed that the dresses she made for me lately, as I was getting ready to go to college, were often cinched at the waist and snug across the breasts and hips. To catch the "right" kind of man someday, I now understood, I was supposed to be a look-but-don't-touch kind of girl.

Without preamble, Mom said while she made the alterations, "Nobody ever explained 'visiting' to me. Aunt Laura and most other folks in the family did not want to remember the past, and for Daddy, even thinking about some issues was not consistent with being a good Christian. Certain matters were not to be discussed, especially between father and daughter. Finally, when I was around your age, I figured out for myself what happened to Mandy."

"What happened?" It had been years since she had promised me an explanation. By this point, I thought I knew the answer, but I was eager to hear the details.

"When Mandy was fifteen or sixteen," Mom said, choosing her words carefully, "Madison Sr. took his pleasure with her. As a result, she gave birth to his daughter Coreen. Then, when Coreen became a young woman, his son did exactly the same thing to her. Coreen might have been a little older."

The sewing machine whirred faster. I knew what Mom was not telling me, and that knowledge was unsettling to both of us. She and I were women too.

When Mom came to deliver the box, some twenty-nine years after the prom, she was still uncomfortable talking about what had happened to Mandy and Coreen. By then, I was married and had my own teenage daughter. I felt awkward talking to Nicole about sex and sexual abuse, but I was clear, and my daughter was tolerant.

My mother, my daughter, and I were far safer than our enslaved ancestors had been, but I knew that for five generations, some of our forebears had been slaves and others had been slaveholders. It was hard to accept that some of our ancestors were victims and that the perpetrators of the crimes against them were family too. Every member of my African-American family, the Other Madisons, I painfully understood, was a product of unwanted sex and incest.

One day during my mother's visit, we sat together in the living room, the translucent curtains muting the gold of the afternoon sun. My mom spoke, and her hands, resting on her lap, looked pale against the bright fabric of her pantsuit. I envisioned Madison

Sr.'s pale hands against Mandy's nearly black skin. White hands touching dark skin are part of my family's history.

"In 1797," Mom said, "Madison Jr. left his political life in Philadelphia and returned to Montpelier. He had represented Virginia at the Continental Congress from 1780 to 1786, played a major role in the writing and ratification of the U.S. Constitution from 1787 to 1789, drafted the Bill of Rights and helped get it ratified in 1791, and served in the U.S. House of Representatives from 1789 to 1797.

But Madison wasn't satisfied. He felt he had nothing that was truly his. As the oldest son, he was first in line to inherit his father's assets — the money, the plantation, the business enterprises, and the slaves. He had married the vivacious widow Dolley Payne Todd three years earlier, but when the couple arrived in Virginia, she was the future president's only 'possession.' He wanted more than that. He wanted children. When he married Dolley, she already had a child from her previous marriage. Madison was happy to serve as father to that child, but he wanted his own. So he imposed his attentions on Coreen. And they had a son."

"Just like that?"

"There must be a story behind it, but nobody knows the details, like when or how many times. Most likely she was ashamed or afraid to tell anybody except perhaps Mandy, her own mother."

"Why Coreen?" I asked. "She was his half sister. He must have known that."

"Daddy, your Gramps, told me that Madison was fond of apple pies, and it was known around the plantation that Coreen's were the best. He saw her walking back and forth between the kitchen and the mansion. He wanted more than her pies. She was pretty and young . . ."

"And his slave," I added.

"The plantation was a close-knit community, but he was away a lot. Maybe he assumed her father was one of the white men working there."

"Assumed? *Wanted to believe* is more likely."

"In either case," Mom continued, "I think emotional need blinded him."

"Or lust." I did not want my mother to protect him.

"Maybe that too," she conceded. "But mainly, I think, he needed to prove to himself he was a man."

"What kind of excuse is that?"

"It's a reason," Mom answered, and it occurred to me that she also needed to protect herself and her pride in the family legacy. She would not, in her way of thinking, dishonor the family.

"But *we* were blamed for the incest, Mom. 'Happy whores' is what the massas called us. To them and most white folks around, we were barely human. They actually believed African women had sex with orangutans!" I said. "And those same white folks also believed that when black women got here and saw white men's pale skin, they couldn't help but use every feminine ploy they could come up with to get next to it. Madison might have thought that way. His buddy Tom certainly believed that nonsense. Jefferson wrote it in a book, probably with his favorite slave, Sally, in bed next to him. Washington might have been a member of the club too. I don't know if George claimed helplessness when faced with the supposed animal powers of black women, but he did father a son with a slave named Venus who didn't even belong to him. Mom, what the most powerful men in America did to the most vulnerable women is sickening. And imagine how those women felt — used, defenseless, angry, degraded."

We sat in silence. I knew what I was saying undermined her reverence for the Founding Fathers, but now that we had started this conversation, I could not stop. I might not have this opportunity again.

"Mom," I said, "saying that Mandy attracted the master's attention because she was such a good worker implies that she seduced him, that he was the victim or, at the very least, that she was responsible for her violation. But in fact, he probably thought she was so eager to get her hands on his white skin and so oversexed, she could not be violated."

"Getting upset doesn't help, Dolly. And you shouldn't think ill of the Madisons. That's just the way it was back then."

"That doesn't make it right, and they knew it. White southerners called what their men did to black women 'trespassing,'" I said. "That meant some man had trespassed on the master's property. Since a master couldn't trespass on his own property, what was it called when *he*, Madison included, went from cabin to cabin?"

"We've always said *visiting*," Mom said.

"What a cowardly euphemism, especially for the actions of a supposedly great man. The result of all this *trespassing* and *visiting* was a lot of mulattos."

Mom nodded. "Some mixed-race children and white children looked so much alike that when a bunch of them were running around a plantation, the only way to tell which ones were slaves was by their clothes. Raggedy—part white. Brand-spanking new—all white." She chuckled. I didn't.

My friend Danielle had told me that Canadians say, "If you shake any family tree, an Indian feather will fall out." I could coin a similar saying for Americans: "If you shake any family tree, a chain will rattle."

═══

The slave community knew that every little black girl was at risk. Did anyone warn these children? Did black girls know that when the time came, there would be little or nothing they could do to protect their own bodies?

One morning several years ago, I took Amtrak from Boston to New York City and then a subway to Harlem to the Schomburg Center for Research in Black Culture. I sat at a long oak table and studied photographs depicting slave life. On page after page, the subjects focused gravely on hoeing rows of soil or picking cotton or tobacco, or they posed rigidly in front of a shack or beside a field. I did not find a single snapshot of children playing.

I returned home the following morning. That evening while I scoured a pot, I thought about my carefree childhood. I'd had two best friends, Patsy and Sheila. We went everywhere together. One afternoon when we were eight or nine years old, we went to a movie theater to see *The Wizard of Oz* for the fifth time. Afterward, at Sheila's house, we figured out how to skip down the Yellow Brick Road just like Dorothy and her companions. All of Sheila's neighbors heard "'We're off to see the wizard, the wonderful Wizard of Oz!'" as the three of us, arm in arm, skipped down her street, singing at the top of our lungs.

Gazing beyond my vague reflection in the darkened window above the sink, I recalled that day on Sheila's street. In place of my childhood friends and me, I saw this:

Three earthbound angels, carried on clouds of dust, frolicked hel-ter-skelter past rows of slave cabins. Skipping along—arms linked, feet dancing, legs flying, heads bobbing—the girls were oblivious to everything except their glee. For them, at that moment, there was no bondage and no past or future, only the untroubled present. I imagined the sound of their joy as they sang in the cadence of an African chant, unaware of the irony of the words, picked up from games played with the master's children.

> Curly Locks! Curly Locks!
> Will you be mine?
> You shall not wash dishes
> Nor feed the swine
> But sit on a cushion
> And sew a fine seam
> And feed upon strawberries
> Sugar and cream.

The girl on the right, the tallest, had fair skin and long, wavy hair. As she skipped and sang, one of her slender hands wafted through the air to the tune's melody and enhanced her fluid movements. This mulatto, when she got a little older, would likely become an obedient house servant, an unobtrusive addition to the décor of the Big House — her voice and joy hushed.

On the left, the clenched hands and stomping feet of a sinewy, dark-skinned girl punctuated the percussion. Kinky hair formed a short, tight braid at the crown of her head, and she had deep-set black eyes, prominent cheekbones, and a broad nose. This young slave's destiny lay among a million rows of tobacco or cotton or rice or wheat or sugarcane or hemp fiber grown for the wealth and security of "the massa" — her voice and joy stifled by exhausting labor.

In the center, supported by her friends, a short, chubby girl with round cheeks and medium brown skin danced, her soft abdomen and round thighs bouncing. Breast buds pressed against her dress. A few years later, this child, sent to lie with one virile black man, then *another, then another, would become a breeder, valued by her owner but pitied by her fellow slaves — her voice and joy reduced to a sigh and a moan.*

Which slave, I wondered, did Massa choose first when he roamed from cabin to cabin? The delicate woman with European features? The sinewy woman who did not tire? Or the one who had already known many men?

I do not recall when I first heard the word *rape*. Certainly not from my mother. But it lurked in the shallowest shadows of history whenever Mom — calling forth a multigenerational memory, long steeped in a thick amalgam of sorrow, fear, resignation, and anger — referred to any abusive white man as "the massa." From slavery to Reconstruction to the Jim Crow era and beyond, black women, especially southern black women, were vulnerable.

In his 1920 essay "The Damnation of Women," W.E.B. Du Bois writes:

> I shall forgive the white South much in its final judgment day: I shall forgive its slavery, for slavery is a world-old habit; I shall forgive its fighting for a well-lost cause, and for remembering that struggle with tender tears; I shall forgive its so-called "pride of race," the passion of its hot blood, and even its dear, old, laughable strutting and posing; but one thing I shall never forgive, neither in this world nor the world to come: its wanton and continued and persistent insulting of the black womanhood which it sought and seeks to prostitute to its lust.

Neither Du Bois nor my mother used the word *rape*. Du Bois spoke out in anger and refused to forgive the South's abuse of black women "to prostitute to its lust." But when Mom said, "That's the way it was then," there was resignation and acceptance behind her words.

In the post–Civil War South, the massa had lost the right to roam freely from cabin to cabin, but the "insulting of the black womanhood" did not end. It was not rare for southern white men to view my grandfather's sisters, my grandmother, my mother, and other black women of their generations as unworthy of respect. Mom might not have allowed herself to use the word *rape* even in her thoughts, but she knew what could happen. She also knew that if she became a victim of this American "tradition," there would be little she could do about it.

What had Mom felt as a young woman living in the South when a white man eyed her tan skin, slender legs, and full breasts? What did he see? A beautiful woman? A black woman? Which made him more predatory?

In 1992, shortly before heading to Virginia to begin my research, I called my mother.

"You've always been so proud of descending from President Madison," I said. "So was Gramps. I don't understand why."

"You don't?"

"No. President Madison and his father were rapists."

"Really?"

"Yes. Rapists. Mom, we've been talking around this for years."

For several moments, neither of us said anything. Although Mom had never called either Madison a rapist, I had assumed she recognized, if only subconsciously, that that's what they were. But now I was not sure. This was the first time she had heard me use that term, but Mom had often said to me, "You shouldn't think ill of them. That's just the way it was." She was defending them from an accusation I had not yet made.

"Maybe they loved them," she finally said now.

"No. The Madisons loved their wives. They *used* their slaves."

"At least the Madisons were accomplished and intelligent, especially the son. He became the president, you know. It wasn't just anyone."

Shortly after that conversation with my mother, I met Nola, an old friend, for breakfast in a tiny storefront café not far from where I worked. We were the only customers. Looking up from the plastic-coated menu, she asked, "How's your research going?"

"Well, right now I'm stuck."

"On what?"

"I don't know what kind of rape Mandy and Coreen had to deal with."

"What *kind* of rape?" Nola dropped her menu onto the Formica table. "Does it matter?"

"I think so. I think it makes a difference whether they were physically attacked or whether they just gave in to the wishes of their masters because they had no choice."

"You shouldn't say '*just* gave in to.'" She sat upright; her eyes focused on mine. "I know *exactly* what rape is," she said. "You *don't* just give in to it."

I looked into my friend's face and saw all of the pain and shame my mother and the generations before her had tried to deny. I also saw anger. I had known this talented, outgoing, attractive woman nearly fifteen years, but I didn't know Nola had been raped. Ashamed of being ignorant and insensitive, I could say only "I'm sorry."

"Don't be. I'm not upset with you," she said, reaching for my hand.

I searched for words that seemed adequate.

"Phil," she said simply.

"Your ex-husband?"

"That's the one," she answered, trying to sound upbeat.

We were sitting by a window, and the white morning light streaming through it washed the color out of her face; she looked like an overexposed photograph. Staring over my shoulder at the street scene behind me, Nola said, "From practically day one of our marriage, he slept with anything and everything in a skirt, but even so" — tears welled up in her eyes — "Phil forced himself on me once, twice, sometimes three times every . . . single . . . day."

I moved my chair to her side of the table and put my arm around her shoulders. The waitress had started toward us across the linoleum floor but turned back to the kitchen.

"We stayed married for over ten years, and it never stopped. Finally, my cousin . . . you've met Emma . . . and my mom and dad came and got the kids and me. Abducted us, actually. Daddy — he had always taken such good care of me until I insisted I could take care of myself — found an apartment for us. I felt like a fool, but he never said anything about my mistakes. And there were many. My judgment when it came to men was worse than bad."

I handed her a packet of tissues, and she wiped her eyes and blew her nose.

"This thing with Phil wasn't about sex," she explained. "This was about power and craziness. He was much bigger than me, and years later, I figured out he was psychotic. I was hurting, and I won't ever be able to forget that, but while we were still married, I thought I had found a way of taking whatever he could dish out.

I would lie there, kneel there, stand there, squat there, sit there, or lean there and think about my beautiful children. I thought about what they were doing in school, in life, what they needed from me, what I could give them. Sex with that crazy man became an out-of-body experience. Even the time he broke the fingers on my right hand—*on purpose,*" she said. "I would become someone else. Anyone else. And I took myself somewhere else. Anywhere else. It wasn't me. I wasn't there when he raped my body."

Nola's eyes, though red with hurt and rage, were strong with self-awareness. Then she smiled. The pain was not behind her, but now she was thriving in a new marriage.

I was surprised at the ease with which she answered my unvoiced question.

"You're wondering why I stayed with Phil so long." Laughing, she said, "I spent a lot of bad hours and a lot of good money on that one myself. Ten years is a lifetime when every day is filled with fear of being murdered. He swore he'd kill me if I left him, and I had no reason not to believe him. When I said the marriage vows, I entered a covenant with God. Then my husband became the father of my children, and I wanted them to grow up in a complete family. He never laid a hand on the kids. I didn't realize how much they were being hurt by what they saw him do to me.

"But I didn't stay just for them. I truly believed that if I just tried harder, if I did things a little better, I could make him love me. And I couldn't give up because that meant defeat, so I baked bread, hand-washed his jeans, served him breakfast in bed. I took care of the children and worked two jobs while he hung out at home, but my husband thought that, as my husband, he deserved more and more and more, everything. We both came to believe that he was omnipotent and that I, failing to meet his needs, most of which neither of us could figure out, had failed as a woman. We believed I deserved to be raped."

I lay in bed that night replaying Nola's words. On the face of it, her circumstances were different from Mandy's and Coreen's. Nola was married, her children would never be sold away from her, and she lived in a home she and her husband could call their own. In that sense, Nola was a free woman. The Thirteenth Amendment promised her freedom, but my friend had been held in bondage.

Lying there listening to the branches of the crabapple tree scrape across my bedroom window, I imagined the massa's "visit":

The dark contour of a well-dressed man appeared in the cabin doorway; his acres of tobacco, his small cotton field, his ironworks, his brandy stills, his mansion, were not visible from her vantage point as she lay on her pallet on the dirt floor.

He stepped into the cabin.

He walked across the small, hot room and stood over her.

Had she known he would come?

Did she recoil, or was she resigned?

What was on his face? Power? Craziness? A smile?

Did he say her name just once, to label her as his possession? Or did he say it over and over, mocking her servitude, her weakness, her helplessness? Did he say her name at all?

Did she say, "Massa," the first syllable pushing from the back of her throat, the second erupting in a hiss?

Did she scream, or did she clench her teeth?

Did he slap her, or did he engulf her in his arms?

Did her mind flee to some safer place, some sweet memory, some pleasant dream?

But . . . what was in his face would not matter.

Whether or how he said her name would not matter.

Whether he broke her fingers or kissed them would not matter.

Whether she screamed and resisted or silently succumbed would not matter.

Whether her mind fled or stayed, aware of his hands on her body, would not matter.

She was his slave.

Mandy

I cried when I saw you. You, my baby girl, were streaked with my blood, the blood of an enslaved woman. You were beautiful, like my mother, though your skin was lighter and more golden, evidence that you would never see the plains surrounding my village back home. Evidence that your father was not the husband I'd dreamed of after my woman-to-be parts had been prepared for him. Evidence that there had been no bridal ceremony and wedding night, only pain that screamed loud in helplessness and louder in hopelessness. When you writhed in my reluctant arms, your back and legs felt rigid, like the trunk and limbs of the tree that had betrayed me. Slick with the water from my womb, your hair burned red, as if the sun had set you on fire in order to cremate your father's sin, and mine. When I tried to kiss your balled-up little cheek, my tears washed over you, salty like the ocean I had lost. You cried, and your scream was terrifying, charged with the battle against life's struggles to come. And I remembered the evening I arrived in this place and the singing I heard in the distance. The melody, I now knew, was a futile lullaby for an enslaved child. Knowing I had brought a slave into the world and that I could not protect you, I bowed my head and cried bitter tears.

12

Sanctuaries

M andy and the other slaves who survived the Middle Passage — not a few of them raped aboard the ships — must have dug deep into themselves to hold on to their humanity. But a solitary strength would likely falter in a place that was determined to destroy it. So slaves drew toward each other. They built communities and created a rich and vibrant culture that was uniquely African, uniquely American, and uniquely southern.

Through reading *Life in Black and White* by Brenda Stevenson, I learned that both the African village and the American slave community, each with a strong sense of kinship, were extended families. Women sat together, chatting and weaving traditional textiles and baskets and cooking yams, gourds, and other West African foods. The women honored childbirth, menarche, courtship, marriage, and death with rituals of joy or solemnity. Everyone, male and female, old and young, stole time to dance to the banjo. Even under the plantation owner's final authority, a couple wanting to marry had to consult their parents, other kin, and other authority figures in the quarters.

Men and women worked together to build shacks that appeared to be little more than cells and transformed them into homes. Dirt floors were swept and walls scrubbed, for this was where a family ate and slept together and where they might find something to think of as their own, no matter how tenuous and transient. While the mansion symbolized wealth and power and a slave cabin poverty and servitude, both structures were homes that offered sanctuary and solidarity.

John Michael Vlach in *Back of the Big House* clarified for me that no matter what their religious tenets were in Africa, most slaves came to believe that the death of Jesus Christ on the cross had saved all of mankind, not just those who were free and white. Otherwise, how could slaves have accepted Christianity so wholeheartedly, and how could the master's religion have become a sustaining force for them?

Slave owners, even those who did not want their slaves to hold meetings of any sort, believed that letting them celebrate Sundays and Christmas was the Christian thing to do. Vlach states: "Religion, in the hands of many slave owners, was an instrument of social control." Some masters hired ministers to preach meekness and submission. I loved the part in Vlach's book about how, after the clergymen left, the slaves held their own services, and "real preachin'" began. "Slaves," Vlach writes, "turned Christian doctrine to their own purposes and created a ritual means by which they could find a spiritual release to compensate for their lack of personal liberty."

As a Christian myself, I was pleased to learn that by adopting and adapting Christianity, slaves found solace and hope in a world that abused them. They accepted Christ not only as their Savior but also as their true master. They celebrated Christ with shouting, jumping, singing, and dancing in ways that expressed their African roots and reminded them that, through faith in God, they could never be truly conquered. They prayed and sang about hope for a better life in this world and the promise of a better life in the next.

As a child, I did not realize that Christianity, along with my family's directive, would guide me through life. During our visit to Texas in 1948, when I was five years old, Mom and I went to see an elderly family friend living on the border of a vast cotton field a few miles outside the town of Elgin. The old woman's skin was dark and wrinkled, and pearly blue-white rims encircled the black irises of her watery eyes. Her shoulders were rounded, but she held her head high, looking intently and expectantly into our faces. She had been a childhood playmate of my grandparents' and, later, an "auntie" to my mother and her brothers when they were children. Now she was alone and craved company. I wondered why the woman, again and again, placed her right hand into her apron pocket and patted it against her thigh.

The field behind her glowed white in the bright midday sun, and I recall thinking that the rows of cotton bushes, too many of them to count, were pretty. But the house was ugly. Laundry hung on the branches of a dead tree nearby. The entire cottage, made of gray, splintery wood, was about the size of my grandparents' kitchen. To reach the sagging porch we had to climb onto a wooden crate, a high step for a five-year-old. A tattered curtain covered the doorway opening into a single, dim room. I do not know whether sharecroppers built this home after the Civil War or whether it once housed human chattel.

When we entered the cabin, I became frightened. It was small, hot, stuffy, and dark. It smelled of mold and bacon grease. A wood-burning stove, a cot, a table, four chairs, a huge tin basin, and a dresser filled the room. I was afraid a piece of ceiling would fall on me or that a mouse might crawl over my feet. The house creaked and groaned. Moths fluttered. In the faint light coming through the single window, I could make out dozens of photographs of dark-skinned people — babies propped up on wooden chairs, boys of varied heights lined up side by side in front of a white church, an old man, the woman's deceased husband, perhaps, standing near a cotton field and grinning at the camera for all his worth. These unframed photos leaned against empty Coca-Cola bottles, vases stuffed with plastic carnations, chipped ceramic cups, a sugar canister, and a small sack of flour. The only picture hanging on the wall was a framed color print of a painting of Christ. His pale face surrounded by flowing, light brown hair, Christ gazed far beyond the room in which He was revered.

The visit seemed long to me but probably lasted less than an hour. I sat silent, but I remember lots of warm, gentle laughter shared between my mother and her former caregiver. Before we departed, the woman loaded us with biscuits, wildberry jam, candied yams, and pickled green beans. As my mother and I were about to step into the car we had borrowed from Gramps, the woman took me by the hand and led me to the back of her house. At first I thought she was leading me to the outhouse, which scared me more than the cabin, but we continued a few yards beyond it. There, in the midst of gray fissured soil, a small garden flourished. Half was green with neat rows of beans, cabbages, collards, peppers, and lettuce. The other half was a chaos of color — purple pansies, orange poppies, yellow marigolds, red petunias, pink zinnias, and more.

While chickens paraded and cackled and hogs rooted and

snorted close by, the woman reached into the pocket of her blue gingham apron. She held her hand there for several moments, smiling with pride at her garden. Slowly, she lifted an old Bible from her pocket and reached for me. With the arm holding the Bible, she held me against her thin breasts. With her other hand, she gently took both of my hands and rubbed them back and forth over the surface of her beloved tome. She looked at me and said, "Jesus saved all us. He protected me an' my chil'ren an' my home, even made this garden grow. Without Him I wouldn't have nothin', wouldn't be nothin.'"

The woman returned the Bible to its place in her apron, gave it a couple of quick pats, and then slowly bent forward to gather a small bouquet. Handing the flowers to me, she said, "Your roots is in the South. You come down here from California, your mommy say, but this is your home. I know'd your gran'ma and gran'pa since I was only a girl. They folk knew my folk — all us just like family. All us knew Jesus. God brought you down here now, go'n bring you back someday. Go'n bring you back home. You'll see."

I did return from time to time, and my grandparents took me to church services that were loud and spirited. Singing and shouting, foot stomping, hand clapping, drum banging, and tambourine rattling could be heard well before the Lord's house, as Gramps called it, came into view. On a rare visit to Texas, when I was ten years old, he woke me up before sunrise so that he could take me to an Easter service. It was being held in a remote woods miles from any town. He did not tell me why we were attending service there. After we'd been driving for two hours, the sky began to turn the pinkish lavender of early morning. Gramps parked the car along a dirt road, and we walked, hand in hand, about half a mile to a small shady knoll. It was an outdoor chapel. The surrounding tall, dense, leafy trees — a mute choir with outspread arms — obscured heaven above, but patches of light played among the broad trunks. Except for the high-pitched drone of insects and birds, the clearing was hushed. Gramps and I found two seats in the wide circle of folding chairs and joined the other worshipers, heads bowed in prayer.

After several moments, without summons, men and women stood up, walked to the center of the circle, and came together in groups of two, three, or four, joining hands or wrapping arms around one another's shoulders to share heartfelt words of advice, consolation, or encouragement. Their voices, ranging from treble to bass, began in sporadic bursts, breaking long moments of silence. But soon, words started to fire off ever more rapidly until they fused and swelled into a symphony of tones, thunderous yet captive inside the thick woods. From within the unbroken chords, joyous "Hallelujah"s, high-pitched wails of sadness, resolute "Amen, brothers; amen, sisters," and screams of "Jesus, Jesus, Jesus!" sprang up and faded into the trees. Then a resonant baritone voice rose from deep within the hum, beginning as a part of the whole but becoming louder and more separate, an epicenter pressing the crying murmur around it into silence.

When everyone was seated again, the preacher told us that when we let Jesus be our source of strength, today's hardship, pain, sorrow, and disappointment would give way to peace and joy tomorrow, if not in this world, then in heaven, our Promised Land. Christ, our Savior, would show us the way. This was how the slaves used to worship, Gramps explained to me.

Later, as an adult, not having grown up in the South made me feel deprived. Though I did not regret missing the terror of living under Jim Crow laws, I was sad that I hadn't had the nurturing environment of a southern black community.

I recall a middle-aged friend telling me about her recent return to rural Alabama: "We went to church every Sunday morning and every Wednesday night. Tuesday, we had prayer meeting, and Saturday morning was Bible-study time. In between, I went visiting family and friends," she said, glowing. "And everywhere I went, someone brought me a plate of food. I ate barbecued ribs, fried chicken, smothered steak, potato salad, macaroni salad, macaroni and cheese, yams, black-eyed peas, corn bread, corn on the cob, collard greens, mustard greens, and turnip greens cooked with onions and ham hocks. Then, when I couldn't eat any more, they brought me pie and cake, every kind you can imagine, under hunks of homemade vanilla ice cream. And full as I was, I ate that too," she added, patting her stomach.

Like the slave fraternity it had once been, the black community in the rural South was an extended family. Food brought neighbors together to comfort the mind and sustain the body. God brought the worshipers together to comfort and sustain the soul.

I had grown up in a strong community of my own, but the closeness of the large, black middle-class society in the San Francisco Bay Area was different from what my friend had experienced in Alabama. The cohesiveness of my California community stemmed from its members' shared desire to keep moving ahead and assure that their children would grow up to preserve and build upon what their parents had achieved. It was assumed we children would attend college and earn our place in middle-class America. Our parents' generation encouraged us to become role models for the next generation, another reason why, in our household, my brother and I heard, again and again, "Always remember — you're a Madison. You come from African slaves and a president."

Now, decades later, I had returned to the South, just as the elderly family friend had predicted. I recognized that she had risen from my subconscious and appeared in my imagination as the old slave who had bathed Mandy. Each introduced a black girl to the American South, a terrifying place filled with terrifying things, but a place that was not without sanctuary.

Mandy

I cried when I saw you. You, my baby girl, were streaked with my blood, the blood of an African woman. You were beautiful, like my mother, though your skin was lighter and more golden, glistening like the morning sun on the plains surrounding my village back home. When you wiggled in my eager arms, your back and legs felt strong, like the trunk and limbs of the tree standing watch over the village where I was loved. Slick with the water from my womb, your hair glowed red as if the sun had set in it. When I kissed your smooth little cheek, it tasted salty, like the ocean that had embraced me. You cried, and the music of your voice rang out, announcing the life inside you. And I remembered the evening I arrived in this place and the singing I heard in the distance. The melody, I now knew, was a soothing lullaby for a beloved child. I threw back my head and cried happy tears.

13

In Search of the President's Son

Historians trying to substantiate past events and genealogists seeking to make family connections and confirm family histories often contend with the accidental or deliberate destruction of documents. Moreover, African Americans combing through archives in search of evidence of their enslaved ancestors have discovered that slave masters often recorded their human property only in inventories and only by number, gender, and approximate age. Slaves owned nothing, not even their names. Because there are no names to search for in many of the surviving records, descendants must find other ways to piece together evidence that a long-deceased relative, though very much alive in family stories, actually existed.

Most devastating to slaves and to the descendants hoping to trace them was that slave owners often tore apart enslaved families. Slaves were expendable property, and selling them was lucrative or, sometimes, expedient. President Madison, who condemned slavery as "a sad blot on our free country" and "a deep-rooted and widespread evil," sold slaves himself. And he allowed his wife to sell his own son, Jim, about whom no records can be found.

But surely there was something somewhere, or so I told myself. In addition to retracing my mother's path to Montpelier, the Orange County Courthouse, and Salt Lake City's Mormon Family History Library, I searched the internet, subscribed to online ancestry databases, called genealogical societies in Massachusetts, Tennessee, Texas, and Virginia. And I hired a genealogist.

No one could find Jim.

I knew I would not be able to uncover a direct link to his father; Madison had no acknowledged offspring. But he had many nieces and nephews. Maybe, I thought, their descendants could lead me to Jim, and maybe from Jim I could find Coreen and Mandy. I asked Dr. Bruce Jackson, a preeminent geneticist who focused on African-American ancestry, to analyze Y-chromosome DNA from cheek swabs from three of my male cousins. I then approached the National Society of Madison Family Descendants about authenticating my family's DNA. We needed Y-chromosome DNA from the acknowledged Madison line. Only one man, I was told, was a direct descendant of any of President Madison's brothers. He agreed to participate in a comparative DNA study, but just then, in June 2007, the *Washington Post* published an article about my quest, "African American Seeks to Prove a Genetic Link to James Madison." Wary of media attention, the man the society had identified changed his mind. I never learned his name.

Dr. Jackson proposed another solution. He was planning a trip to England for his own research, and he offered to hire British genealogist Ian Marson to locate a descendant of John Maddison, the president's great-great-grandfather. The hope was to find an English successor who, not stigmatized by the history of slavery in America, would be willing to allow his DNA to be compared with my cousins'.

However, Marson was not able to find a living male in Maddison's family line.

I put my search for a genetic connection on hold and went back to the archives. I started looking for clues that might tell me something about Jim's life after he was sold. I hoped to find out who had bought him from Dolley or who had taken him to Tennessee. But the obstacles continued to pile up. Many records, if they'd ever existed, had been lost in a surprising number of courthouse fires.

There were small fires as well. The former president, while organizing his papers, selected missives and documents for Dolley, his closest confidante, to burn. Madison was likely trying to protect his correspondents' privacy and shield them from embarrassment. But I can't help thinking he also hoped to safeguard his legacy.

In the Madisons' era, it was not uncommon for well-to-do people to request their private letters to be burned after their death. I learned that Dolley had instructed her niece Anna Payne to go through her personal papers and decide which ones should be destroyed. Dolley was most likely protecting her own privacy.

The actions of John Payne Todd, Dolley's son by her first marriage, made matters even more difficult for me. He hadn't burned Madison's papers, but he had pilfered and sold many of them. Payne, as he was called, was driven by his need to pay off gambling debts.

After Payne's death, in 1892, items from his estate, Toddsberth, were put up for sale in order to cover his unpaid bills. Prior to the sale, the sheriff and a county justice of the peace went through the Madison papers still in Payne's files and discarded or incinerated most of them. Only documents with monetary value were spared.

Not all of Dolley's papers were burned. She left behind a small gem of little-known information about a celebration at the end of the War of 1812: Dolley had indeed ordered slaves to stand around the room holding torches. Few scholars knew this detail, but for two centuries, generations of *griots* in my family had passed down the story of Jim holding a rushlight at that party. Dolley Madison biographer Catherine Allgor writes: "Dolley may have lost her silver, mirrors and lamps, but she supplied the drama of light for one party by stationing enslaved men throughout the house with pine torches."

The rushlight story corroborates our family's oral history of Jim. (It has always bothered me that I could not find record of Victoria, the young white woman whose love for Jim resulted in his being sold. But several of James and Dolley's nieces — some whose names were not noted — spent long periods at Montpelier.)

To shore up my optimism, I turned to two of my cousins, Sean Harley and Jimmy Madison. They had been searching archives and online resources for evidence of our ancestors for many years. Sean, then in his thirties, had engaged in family genealogy since junior high. His great-great-grandfather was Giles Madison, one of Jim's grandsons. My great-grandfather Mack Madison was one of Giles's brothers. Giles and Mack married sisters, Fanny and Martha Murchison Strain. Sean and I enjoyed the complexity of our kinship: we were double third cousins once removed.

My third cousin Jimmy, a descendant of Charles Madison, another of Jim's grandsons, began his research after he heard my mother's oral-history presentation at a family reunion. Jimmy decided to devote his retirement years to finding more family documents.

Sean, Jimmy, and I joined forces and pooled our information. Looking for clues about the lives of our enslaved ancestors, we pored over census data, bills of sale, wills, lists of taxable property, and whatever else we could find. From oral history, we knew that one of Jim's sons was our ancestor Emanuel, a man we hoped would lead us to his father.

Though we could not find out how, when, or where Emanuel was acquired, we knew from tax records that by the 1820s, Emanuel lived in Tennessee as the property of Jeptha Billingsley (1780– 1863).

When Sean and Jimmy found documentation of a free black man named Shadrack Madison, more clues emerged. We learned that until Shadrack gained his freedom, (probably before 1816 but not recorded until 1817) he was the property of another member of the Billingsley family, Jeptha's father, Samuel (1747–1816).

Were there more clues that suggested a connection between the two African-American men? The archival path linking them was filled with gaps, requiring us to search for more hints, make inferences, and try to turn the scattered pieces of evidence into a coherent story. The next clue came from Montpelier. Jim, according to our family stories, was born on the plantation around 1792. Shadrack, according to census data, was born in Virginia that same year, and Montpelier records (which included shoe sizes for slaves) from 1782 to 1786 list the uncommon name Shadrack among property inventories. My cousins and I conjectured that the emancipated man may have been the namesake of the older enslaved man.

Another link between Emanuel and Shadrack came from records in Gibson County, Tennessee. Both men lived there for some twenty years roughly from 1828 to 1848.

County tax records showed that in the 1820s and 1830s, Jeptha Billingsley owned one black poll (taxable slave). In 1834, Jeptha purchased a wife for that poll. The bill of sale, executed in Gibson County, identifies the male slave as Manuel (Emanuel).

Shadrack arrived in Gibson County in 1828. For at least eleven years prior, he had lived in Bledsoe County. In 1827, only one year before his move, he purchased land from Mary Billingsley, Jeptha's mother. Abruptly, Shadrack sold the land and left his wife and children behind. (He later purchased his family and brought them to Gibson County.)

In 1848, several events that might have linked Emanuel and Shadrack occurred. That year was significant for both men. Emanuel and his family were uprooted from Tennessee and taken to Texas, and Shadrack sold two parcels of land in Gibson County. By 1850, according to the national census, he and his family were residing in White County, Illinois. He had been a free man in a slave state for more than thirty years, and now, it seemed, he no longer felt compelled to stay there. My cousins and I surmised that Shadrack might have remained in Gibson County in order to live close to Emanuel and his family, but once they were gone, Shadrack moved his own family to a free state.

The final, and most significant, clue that there was a close connection between the two men is their surname.

When they were emancipated — Shadrack by a legal document recorded in 1817 and Emanuel by the Emancipation Proclamation in 1863 — neither man took the last name of his owner, Billingsley.

Instead, both took the name that I believe was chosen to honor their family credo: Madison.

As soon as she learned Jim had been sold, Coreen told him to remember that name. It could be a tool, she hoped, that they might be able to use to find each other . . . someday. Now, generations later, the Madison name is, once again, a tool. My cousins Sean and Jimmy and I, unlike innumerable other descendants of slaves, are fortunate to have a name with which to begin to weave a life history. The documentation is scattered and circumstantial, but that is all that is available to us. We have no choice but to accept uncertainty until we can gather more evidence . . . someday.

Could it be that Shadrack Madison was Emanuel Madison's father and Coreen's son? Could it be that Shadrack Madison was our long-lost Jim?

14

Elizabeth

While looking for Jim, I had to rely on circumstantial evidence and conjecture. Not so for my search for my great-great-grandmother Elizabeth, Emanuel's wife. One of my favorite photographs, the oldest in the box, has always been the one of Elizabeth Madison. (On the back of the picture, someone wrote *Ma Madison,* into which I read deep respect and affection.) In the cracked, fading image she is an elderly lady seated on a chair in front of a picket fence and a tall tree. She is a free woman in this photo, but Elizabeth had been a slave most of her life. Sitting alone, she gazes down at her hands resting on the folds of her skirt.

Of all the family stories, Mom's favorite was about Elizabeth's grandmother Katie. Here is how my mother told it: In the French-speaking Senegambia region of Africa sometime during the mid-1700s, Katie's mother, a beautiful African woman, married a French sailor. Over several years, they had three children. Katie was the oldest. When their father died at sea, a well-to-do French merchant asked their widowed mother to come, with her children, to what is now Port Harcourt, Nigeria. He needed someone to clean and cook for himself and his wife.

Once the family had settled in, Katie's mother started working for the merchant. Each day, he sent a wagon to make sure she traveled safely to his house. Katie, by then in her early teens, had to remain home to watch her younger brother and sister. Whenever her mother left their house, she warned Katie to stay inside with all the doors and windows shut tight and locked. Slave catchers were everywhere.

One bright, sunny day, Katie had a few coins to spend. And she was restless, cooped up in a hot, dark house. Her brother and sister, Katie figured, could not get into much mischief if she left them unattended for just a few moments. To increase the likelihood they wouldn't tell her mother she had gone out, Katie planned to bring back a surprise treat for them.

Katie hurried to the market, enjoying her adventure and wondering why her mother made such a big fuss about leaving the house. What could go wrong on such a nice day?

Katie bought a small cake to share with her siblings, then headed home. Suddenly, as she approached the footbridge, two black men stepped in front of her. She screamed and tried to run away, but one man grabbed her wrists, and the other grabbed her feet. The town constable, who had known Katie since her arrival in Port Harcourt a couple of years earlier, saw what was happening, but he didn't do anything. Most likely, he had been paid not to interfere.

It wasn't long before she found herself in the hull of a slave ship, but it seemed an eternity before the ship reached land. Katie disembarked on the east coast of Florida, where a Mr. Edward Jackson purchased her for his plantation in Pensacola.

At first, things went well for Katie. She was chosen to be Mrs. Jackson's personal servant. The mistress enjoyed showing off her French-speaking slave and taught Katie to do dainty domestic work: sewing, embroidery, and making fancy cakes and cookies. Katie had never been in a house with so many stairs, and she liked the feel of her braids bouncing against her back as she ran up and down the steps.

Katie learned English quickly, but she pretended not to understand it. She reasoned that if she did not want to do something or if she made a mistake, she had an excuse. And Katie loved to listen in on Mr. and Mrs. Jackson's conversations.

Every summer, a carnival came to town. On the one day set aside for slaves and free blacks, Toby, a field slave from Togo who worked on one of Jackson's distant plantations, came to take Katie to the fair. In her excitement, she became careless and spoke to Toby in English. Mr. Jackson, standing nearby, overheard her and realized Katie had deceived him. Enraged that a slave had made a fool of him, he gathered his family and every slave at the front of the mansion. He grabbed a switch, threw Katie onto the veranda, and beat her bloody.

It took several months for Jackson to get over his anger at Katie, but when she and Toby finally found the nerve to approach Jackson to ask if they could "jump the broom," he agreed to let them become husband and wife. But he did not allow them to live together, which made Toby Katie's "abroad husband." He could see her only on Sundays and a few nights now and then, whenever he could sneak away from his quarters on Jackson's other plantation. Within a year, they had a daughter, Lilly.

Lilly grew up to be a fine young woman, but quite a bit shyer than her mother. One morning, Mr. Jackson suddenly became ill, and he sent Lilly to get the doctor. On the way back, about a mile from the plantation, she heard footsteps behind her. She kept walking, her head down.

"You're pretty," a man said.

She walked faster. He kept pace with her.

"You're real pretty," he said.

Lilly realized the man was not going to go away and that she probably could not outrun him. She stopped and turned to look him in the face. He took a step back and lowered his head. His hat fell off. As he bent forward to pick it up, his black braids, as long as her mother's now gray ones, fell over his shoulders.

"My name's John Quail," he said. "I'm Choctaw."

"I'm Lilly. I belong to Massa Jackson."

They fell in love that very moment.

The doctor came, and Mr. Jackson's health improved. He gave permission for Lilly and John to jump the broom, but he would not write up free papers for the bride.

Jackson's health declined again, and within three years, he was dead.

Mrs. Jackson was surprised to learn her husband had left behind a considerable number of debts. To settle them, she sold several slaves, including Lilly and her three-year-old daughter, Elizabeth. It happened while John Quail was away delivering seed to a nearby farm. For months, he searched everywhere and asked everyone about his wife and child, but no one could, or would, help an Indian man looking for a pair of slaves. John Quail never saw his wife and child again.

Lilly, like Jim, has been lost to history, but her daughter, Elizabeth, the woman in my favorite photograph, became the property of Augustus King in Gibson County, Tennessee, and in 1834 King sold her to Jeptha Billingsley. My cousins and I were elated when Sean found the sales agreement in which Billingsley purchased a female slave, Betsey, from King as a mate for his male slave Manuel.

Betsey was my great-great-grandmother Elizabeth, Katie's granddaughter. Manuel was my great-great-grandfather Emanuel, Jim's son.

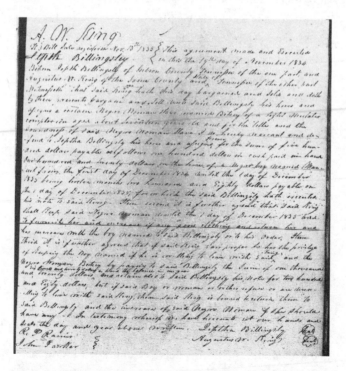

Elizabeth ("Betsey") Madison's bill of sale, 1834

It took several people, including a lawyer, to decipher the faded, handwritten text of that document.

> Bill of sale Registered November 13th, 1835. This agreement made and executed on this the 19th day of November 1834 between Jeptha Billingsley of Gibson County Tennessee of the one part, and Augustus W. King of the same county and state of Tennessee of the other part willful forth that said King this day bargained and sold and delivered then present bargain and sold unto said Billingsley his heirs and assigns a certain Negro woman slave named Betsey of a light mulatto complexion about seventeen years old and for this title and the soundness of said Negro woman slave I do hereby warrant and defend to Jeptha Billingsley his heirs and offerings forever for the sum five hundred dollars payable as followed one hundred dollars in cash paid in hand one hundred and twenty dollars on hire of a Negro boy named Manuel from the 1 day of December 1834 until the 1 day of December 1835 being therefore twelve months two hundred and eighty dollars payable on the 1 day of December 1835, for which the said Billingsley hath executed his note to said King Item second it is further agreed that said King shall keep the Negro woman until the 1 day of December 1835 and to furnish her and increases if any with good clothing and return her and any increases with the boy named to said Billingsley or to his order. Item third it is further agreed that if said King shall prefer he has the privilege of keeping the boy named if he is willing to live with said King and the Negro woman Betsey by paying to said Billingsley the sum of one thousand and twenty dollars six hundred and twenty dollars cash in hand and the balance within one year and return back to said Billingsley his note for two hundred and

eighty dollars, but if said Boy or woman or either refuse or unwilling to live with said King then said King is bound to return them to said Billingsley and the increases of said Negro Woman if she should have any. In testimony where of my hand here unto set our hands and seal the day and year above witnessed

> R. P. Raines
> Jeptha Billingsley
> John Parker
> Augustus W. King

In 2011, when I received a copy of this transaction from my cousins, the consideration the slave masters gave to their property surprised me. King had agreed to "furnish [Elizabeth] and increases if any with good clothing." Furthermore, at the end of one year, the slaves could decide for themselves whether they wanted to live with King or Billingsley and, more astounding, whether they wanted to live with each other. Elizabeth and Emanuel chose to stay together; thus, this document is tantamount to a marriage certificate, a rare find for descendants of African-American slaves. I held the bill of sale to my chest and closed my eyes. Though still someone's property, my ancestors, contrary to the standards of that time and place, had been treated like human beings.

According to various records, Elizabeth and Emanuel had at least twelve children, some born in Tennessee, some in Texas. Most were boys. An 1842 list includes a girl, Manda, but her name does not appear again. Property data reveal that the remainder of the family lived in Tennessee until 1848, when Jeptha Billingsley sent them to the town of Cedar Creek in Bastrop County, Texas.

Texas had been admitted to the Union as a slave state in 1845, and Jeptha Billingsley decided to join the throngs of settlers surging into the region. Texas and other cotton-growing states were the "New South." The "Old South," which included James Madison's beloved Virginia, was ailing. Cotton was in much higher demand than tobacco. Jeptha's son Jesse had moved from Tennessee to Texas in 1834 and later received some twenty-two hundred acres for his leadership in the military. He allotted some of his land to his father, and Jeptha needed slaves to work it. Mom often told a story about what happened next.

"One morning," she would begin, "Emanuel, Elizabeth, and their four oldest sons, Shelby, Mack, Henry, and Giles, were about to start out for the cotton field. A white man they had never seen before appeared in their cabin doorway. In the dim light, he peered around the cabin, glancing at four-year-old Charles asleep on his pallet. 'Wake your young'un. Get your things together.'

"'What for?' Emanuel asked.

"'Y'all are heading out. Mr. Jeptha sent me to drive you to Texas, so hurry up. And no back talk.'

"Before Emanuel and Elizabeth could move, the driver roused the sleeping child, carried him out of the cabin, and put him in an open wagon. Afraid they would never see Charles again, Emanuel, Elizabeth, and the older boys—with no time to gather supplies, food, or clothing—ran to the wagon and jumped on. The driver and a man carrying a shotgun tied ropes around Emanuel's wrists and clamped chains around the ankles of everyone except Charles.

"It was midsummer, and the sun was too hot for them to travel by day, so, though it was dangerous, they journeyed at night. Thieves hid along roads, waiting to ambush and abduct slaves. Throughout the trip, the guard held the family at gunpoint so that they would not try to escape. When Charles cried from hunger and fear as they rode through woods so thick no moonlight shone through, the guard shoved the gun into Elizabeth's chest and whispered harshly, 'If you want everybody to get to Texas, you better shut that boy up.' Charles learned to cry silently.

"The three-week trip seemed endless. At one point, a wheel came off the wagon, and it took two days for the men to find another one. Finally, the Madisons arrived in Cedar Creek. They were starving and dehydrated. Blood oozed from Emanuel's wrists. The boys sat listlessly on the wagon floor. Nevertheless, Elizabeth felt blessed her family was alive and together. Barely able to remember her own father and knowing that many slaves, including her husband, had lost their loved ones forever, Elizabeth would marvel at her good fortune for the rest of her life. She said over and over, 'God is good. God is amazing.'

"Jeptha might have done a good deed by keeping the family together, but it was to his advantage to acquire six healthy males and a fertile woman who were at home in a cotton field. He was horrified to find them in such poor condition when they arrived in Cedar Creek, covered with dirt, too weak to walk, and insects crawling through their hair. His investment was in jeopardy."

The family had left Tennessee without adequate food and water, and no one had given any thought to their personal needs. Years later, Elizabeth described how embarrassed she was that her boys had seen her during her "time of the month" and how helpless she felt when the guard and driver laughed. Though not being able to hide the flow of menstrual blood is a small problem in the face of all the hardships and degradations Elizabeth endured as a slave, Mom and I, as women, felt her humiliation.

In the cherished photograph, Elizabeth's hands are empty, and she is alone, without Emanuel or any of her many children. I never met my great-great-grandmother, but in my childhood, I loved that picture, imagining myself climbing onto her lap and placing my hands in hers. As an adult, I see the gentle strength in hands that picked cotton, created a home, and nurtured children. Her hands had purpose. While Mandy was the mother of our African-American family, Elizabeth was the first of our ancestors whom we could actually see. When Elizabeth sat for her photograph, she left a gift: an image of resolution, love, and faith in God. I keep her picture with me at all times.

Elizabeth Madison (1815–1911)

15

Free!

On January 13, 1857, Jeptha Billingsley granted his son Jesse, a former senator representing Bastrop County in the Texas Legislature, legal charge of my enslaved ancestors. One of the items in the cardboard box my mother brought to me is the deed that bound my family to Jesse. It reads

> Jeptha Billingsley . . . is the bonafide and actual owner in his separate right and property of the following named negroes slaves for life, to wit: A Negro man of dark complexion named Emanuel aged fifty two years and his wife, a woman of yellow complexion named Betsey aged about thirty seven years and their eight children all boys of the following description to wit: Shelby of copper color age about twenty two years, Mack of the same color aged about nineteen years, Henry same color, aged about seventeen years, Giles of same color aged about fifteen years, Young of same color aged about thirteen years, Charles of same color aged about eleven years, James of same color aged about nine years, John of same color aged about four years . . . Now the

party of the first part, the said Jeptha Billingsley, doth for and in consideration of the sum of nine thousand dollars sell transfer and deliver to the parties of the second part, the said Jesse Billingsley . . . all of the aforesaid Negro slaves . . . into the complete possession and controle of the party of the second part.

Jesse, born in Rutherford County, Tennessee on October 10, 1810, was eccentric, a character straight out of the legendary Wild West. He was a renowned Indian fighter, and from 1836 to 1837, he served in the House of Representatives of the First Congress of the Republic of Texas. During this time, he reputedly wore a buckskin suit taken from a captured Indian and at night slept on the floor of the state capitol building. He claimed that he'd supported eighty men on the frontier with wild game and clothed his troops in the skins of animals they had killed. Jesse boasted of his company of rangers: "We were only chargeable to the government for one sack of coffee and one sack of salt." Davy Crockett, a close neighbor in Tennessee, was his hero.

In the battle at San Jacinto in April 1836 — a decade before Emanuel and his family arrived in Texas — Captain Jesse Billingsley commanded the first company to meet the Mexican forces. During the fighting, he suffered a wound that would end up crippling his left hand for life. Evoking Crockett's death at the Alamo the previous month, Jesse coined the famous battle cry "Remember the Alamo!"*

---

* Harry Alexander Davis's *The Billingsleys of America* attributes the cry "Remember the Alamo" to Jesse Billingsley, but other sources attribute it to General Sam Houston.

In the cardboard box is a copy of the November 1939 issue of *Frontier Times,* which includes the story "Captain Jesse Billingsley: A Texas Patriot." At the lower left corner of the magazine article is a photograph of a thin-faced man with bushy eyebrows, unruly dark hair, and a white-streaked beard. His eyes are fierce and direct, his formal attire — dark coat, high collar, and wide bow tie — incongruous with his tempestuous face and hair. When he died, on October 1, 1880, the Billingsley family discovered a document stating his desire to be buried with his horse Gofer and his parrot in the front yard of his house in McDade, Texas. Jesse's family complied with his wishes. In 1929, he was reinterred in the Texas State Cemetery in Austin.

Jesse Billingsley (1810–1880)

The box of family memorabilia now in my care also contains a copy of an 1860 slave census, the last one before the Emancipation Proclamation in 1863. Under Jesse Billingsley's name are the ages and genders, but not the names, of Elizabeth and Emanuel and their growing family. The parents, nine boys, and a baby girl are on the list. From their ages, we can tell that the five oldest of those boys were Shelby, Mack, Henry, Giles, and Charles, the ones who had traveled from Tennessee with Elizabeth and Emanuel. The other children were born in Texas. Three of the younger boys are James, Young, and John. What happened to the ninth boy and the baby girl is not known; they likely died young or were sold away.

1860 census (names of slaves added by my mother)

Despite bondage, Elizabeth and Emanuel tried to instill in their children a sense of pride. Emanuel's father, Jim, had taught him to read, and Emanuel, with the approval of Jesse Billingsley, taught his own sons this now-forbidden skill. After Nat Turner, a literate preacher, and his band of warrior-slaves murdered some sixty white men, women, and children in Southampton County, Virginia, in 1831, most Southern states had adopted laws making slave literacy illegal. Nonetheless, Emanuel's family read the Bible every day. The boys learned the teachings of Jesus; they learned to be faithful to God and be dependable, trustworthy, and hard-working. And they were constantly reminded of their great-grandmother's plea to remember their family name. They were Madisons, and they were together. Still, the threat of being sold and separated hung over them.

Four years after Jesse acquired my ancestors, Abraham Lincoln became president. He took office in March 1861, and one month later, the North and the South were embroiled in a war that would last more than four years, take the lives of more than seven hundred thousand Americans, and free more than four million slaves. What began as a disagreement over the expansion of slavery into the Western Territories finished in a face-to-face bloody battle to end the "peculiar institution" throughout the nation.

The Civil War came as a terrible shock to the citizens of Bastrop County, Texas. Most of the farmers owned only small numbers of slaves, if any, and favored abolition over war. Led by Jesse Billingsley, the county voted against secession, arguing that Texas had been admitted to the Union less than twenty years earlier, and its status as a state should not be cast off so soon. Other counties rejected this reasoning. Bastrop had no choice but to secede with the rest of Texas.

"Little Dixie"—Oklahoma, Louisiana, Arkansas, and Texas —was slavery's last stronghold. Slave owners sought to hide their most valuable property from the Union army. During the last two years of the war, more than 180,000 slaves from the Deep South were herded into eastern Texas for safekeeping. The result was an abundance of free labor on an abundance of fertile soil in a cotton-growing state. Perhaps Billingsley hid Emanuel and his family among the slaves from other states, but it seems more likely to me that they stayed right where they were, in the mid-Texas countryside of Cedar Creek, doing what they had been brought there to do: sow, reap, and bale cotton.

In 1863, the Union was desperate for cotton, so President Lincoln allowed federal agents to buy cotton from the enemy. While the official Confederate policy was that the South would sever all commercial contact with the North, Texan planters, probably including Jesse Billingsley, betrayed the South by selling Northerners cotton. Northerners betrayed the North by giving Southerners money. In pursuit of cotton and money, Northern and Southern profiteers prolonged the war by at least a year, costing tens of thousands more lives.

Initially, the Emancipation Proclamation, issued January 1, 1863, freed slaves in only the ten Confederate states in revolt, which included Texas. But Lincoln had no jurisdiction in the Confederacy, so fighting continued. One of the proclamation's ultimate powers grew out of its provision that black men could serve in the Union army and navy, a change of policy that contributed significantly to the North's eventual victory. Nearly 200,000 black soldiers, in segregated units supervised by the Bureau of Colored Troops, fought in approximately two hundred skirmishes. Another 130,000 enslaved men and women supported the Northern cause as laborers, construction workers, caretakers of horses and mules, and cooks. Enslaved people were not mere spectators, as it often seems in textbooks.

Slaves throughout the South knew about Lincoln's proclamation, but the Confederacy and its struggle to maintain slavery persisted. Finally, on April 9, 1865, General Robert E. Lee surrendered to General Ulysses S. Grant, who had taken full advantage of the willingness of African Americans to fight for their personal sovereignty. In effect, by helping to save the Union, slaves freed themselves.

The Madisons knew they were free but were cautious about exercising that freedom for two and a half years, until June 19, 1865, "Juneteenth," when General Gordon Granger arrived in Galveston, Texas, with federal troops. Then, as my mother was fond of recounting, the eight boys, along with almost everyone else around, whooped and hollered for hours.

"Emanuel smiled at Elizabeth," Mom liked to say. "He took her hand and led her to a tree stump on the far edge of the cotton field. He sat down, then gently pulled her onto his lap and wrapped her in his arms. His voice tremulous with love, joy, and anticipation, he said, 'I've always loved you. I've always loved our boys. But now I can love you and them even more. Nobody can take you away from me. Nobody can take our boys away from us. We're free to love each other as much as we want to. Only God is our master now.'"

The family had come closer to separation than Emanuel knew; Jeptha's 1862 will carved up the Madison family among the Billingsley heirs, but Jeptha died on October 25, 1863, nearly ten months after President Lincoln issued the Emancipation Proclamation on January 1.

For the first time, for all African-American families, the threat of being sold and separated was gone. The Madison family saying became a reminder that five generations of enslavement had not destroyed them. They could have taken Billingsley's name, but Jim's promise to his mother had instilled in them who they were.

For them, *Madison* meant family, and they were proud the name had belonged to a president. Jim, Emanuel, and Emanuel's children had known they were Madisons, but once free, they could officially claim the name.

Gramps described emancipation as a time of great jubilation. The joy of freedom, full of expectation and hope, was different from any joy the former slaves had ever known. By law, they were human beings with rights, not someone's property. Throughout the South, freed slaves sang and danced. They plucked banjos and shook tambourines. From the tip of Florida to the westernmost edge of Texas to the nation's capital, celebrations erupted like firecrackers and were just as loud.

But Reconstruction, 1865 to 1877, was a time of national chaos, instability, uncertainty, and vicious racial retaliation. There was no clear plan for rebuilding the war-torn South physically, socially, economically, or politically, no clear plan for reuniting the South with the Union, and no clear plan for the welfare of millions of homeless, penniless freed slaves. The proposed allotment from the Freedmen's Bureau of "forty acres and a mule" became a myth. No one would sell land to former bondsmen, even if they had money. There was no choice but to rent, at exorbitant prices, from former slaveholders.

Having dreamed of freedom throughout their lives, the Madisons, like nearly every other enslaved family, thought being emancipated would solve their problems. What they soon learned was that the American dream would not be theirs without a fight. The ideals of the Declaration of Independence, the Constitution, the Bill of Rights, and even the new amendments written specifically on their behalf betrayed them. Southern whites — angered by the outcome of the war, hanging on to a past that had taught them they were superior, and terrified that their future was now threatened — did whatever they could to turn the slaves' emancipation into a state of destitution and fear. The sharecropping system was impoverishing. The Black Codes, laws that restricted the lives and livelihood of former slaves, were oppressive. And the Fourteenth and Fifteenth Amendments, meant to guarantee civil rights and the right to vote to freedmen, were unenforceable. Black people, though no longer enslaved, were not in fact free.

When freedmen voted in large numbers, putting black Republicans in office on local, state, and federal levels in the 1866 congressional elections, white Democrats resolved that black people would not vote again. The South had lost the war, but belief in the Confederacy — which had lured white men and boys to injury and death in order to fight for slavery — persisted and erupted into intense hatred for blacks. Terrorist organizations such as the Ku Klux Klan, founded in 1866 by veterans of the Confederate army, threatened to beat, burn, or lynch any black man who dared to vote. Using physical and economic coercion, white supremacists regained control of the South, forcing blacks to give up their constitutional right. Blacks heard their employers say, "If you vote, don't bother coming back to work."

Nevertheless, according to our family stories, soon after the eight Madison brothers and their wives and children heard the Emancipation Proclamation, they sought ways to take advantage of their newly gained freedom. Emanuel told his sons, "We've come far, but we've far to go." He said, "Your great-grandfather was president of this country. You can do great things too, so make something of yourselves, now that you have the chance. Tell your children, and tell them to tell *their* children: Always remember — you're a Madison. You come from a president."

The family saying was now a source of inspiration, and the work of living up to it began. From then on, my freed ancestors knew that with every step forward there would be at least one step back. Freedom was a responsibility, and it was up to them to see to it that their families were clothed and fed. It was up to them to find a way to get their children educated. Every decision was theirs to make; the consequences, good or bad, were on their shoulders.

Henry Green Madison (1840–1912)

My great-uncle Henry, the third oldest of the boys, was twenty-three, married, and had a child when emancipation came. Right away, he decided a farm was not the place for him. He wanted to educate his children and make a better life for his family. His wife, Louisa, and daughter, Elizabeth, stayed with his parents in Cedar Creek for a year while Henry traveled the twenty-five miles to Austin to learn carpentry. He returned to Cedar Creek every weekend to have Giles and Mack help him cut logs. Once the three brothers had loaded the logs onto Henry's horse-drawn wagon, he hauled them to Austin to build a cabin. When it was finished, Henry brought his family home.

More than one hundred years later, a wrecking crew demolished an 1886 frame house at 807 East Eleventh Street in Austin and discovered an intact rough-hewn cedar cabin in the middle of the rubble. City records revealed that the log cabin was built around 1863 on the location registered as the homestead of Henry G. Madison. Henry had not destroyed the first home he could call his own; instead, he kept it as a central living space within his relatively grander house. In the summer of 1968, the city moved the log cabin to Rosewood Park, a recreation center in one of Austin's black neighborhoods, where it stands today, a monument to the substantial roles slaves and ex-slaves played in American history.

Thirty years later, in 1998, I flew to Austin to see the state historical site I had heard about since childhood, my great-uncle Henry's cabin. The early-morning sky was dark when I left the hotel, and the rising sun was still low when I parked the car and climbed a grassy hill. Behind me, teenage boys kicked a soccer ball and shouted to one another in Spanish. At first, I could not see the cabin, but I sensed I was headed in the right direction. As I reached the crest of the hill, the sun glinted through a sparse grove. Then I saw it. It seemed to have appeared suddenly in the open space in front of the trees. The cabin was tiny, approximately twelve feet on each side. I was astounded that a family of two adults and seven children — six of whom had entered the world in that cabin — could live in such a small place, and for twenty-one years.

I wanted to go inside, but no one was around to unlock the door. While trying to peek through a crack in one of the crude wooden shutters Henry had made, I felt the rough texture of the logs he had cut and hauled and that had remained in place for more than a century. I walked around the cabin, touching and stroking the wood Henry had cut, the coarse mortar he had mixed, the thick nails he had pounded to build a home for his family. I felt his strength and determination against my hands. That power had not diminished over the decades or with the passing generations or under the pressures of an ever-changing world. That power was right there — under my fingertips.

Henry Green Madison's cabin in Rosewood Park, Austin, Texas

Mom often talked about Henry's brother Mack, my great-grand-father. When emancipation came, he was twenty-five and married to Martha, the mulatto daughter of Mahala Murchison Strain, the first Negro in Austin.

Mahala Murchison (1824–1912)

Mack, too, had a plan for his freedom. He wanted to remain near his aging parents, and Cedar Creek, he decided, was not a bad place to live, even though the soil was dry and gravelly and the weather hot and humid. He liked the rolling hills and the thick groves of giant cedars that grew along the creek. Choctaw, Tonkawa, Delaware, and Comanche Indians still roamed the country, but they had suffered massive defeats, so raids and battles with settlers were much less frequent.

Another reason Mack stayed was that he had become Jesse Billingsley's right-hand man. Neighbors were shocked when Billingsley made Mack his business manager. Mack took cotton to the gin, negotiated deals, and even handled the money. Mom often exclaimed, "Imagine that, a colored man touching a white man's money!" What outraged the neighbors was that Jesse advocated for slave literacy. This outrage was at least one of the reasons that Billingsley, though a war hero, did not get reelected to Congress.

Mack and Giles were tenant farmers, "halvers," under an agreement to give half of their income from the farm to Billingsley. Little by little, Mack put away money, and after nine years, in 1874, he had $192 in gold, enough to buy a ninety-six-acre farm. For the first time in his life, Mack owned something, and not just anything — he owned land. Security, survival, and true independence were within reach. Once the deed was in his hands, he left Billingsley. The very next year Mack sold the farm for $394 and purchased two hundred acres for $400. Copies of the deeds to both properties are among the treasures in the box.

Mom enjoyed describing how proud her father looked whenever he entered his sister Laura's living room. He went there frequently and spent hours looking at photographs of his parents. Mack and Martha had been slaves, but they were Gramps's king and queen. To express his pride in descending not just from a president but from slaves as well, Gramps was the one who added two crucial words to the family credo: African slaves. Since then, we say: "Always remember — you're a Madison. You come from African slaves and a president."

Coreen's words were vital to my family both when they were listed as someone's possessions and, later, when they could take a name and then make a name for themselves. For my freed ancestors, the admonition was an inspiration for pride and achievement — for themselves and for the generations to come.

16

Gramps

During Reconstruction, life was tense for everyone in Texas, and dangerous for African Americans. Local governments parceled off plantations. Railroad lines carved up the landscape. In Cedar Creek, after General Granger's troops arrived to enforce Lincoln's proclamation in Texas, some forty black men formed the Negro Loyal League. Its purpose was to protect the fragile status of the African-American community. Every Saturday night, the men met at the Cedar Creek Store and performed drills up and down the road, carrying whatever weapons, including sticks and rocks, they could get their hands on.

One evening as Mack was on his way to a league meeting, Peter Murchison, Mack's white cousin through his marriage to Martha, stopped him on a bridge, pointed a gun at his chest, and refused to let him pass. Peter and Mack had been friends for many years, but that evening, Mack, like every other black man in town, was seen as a threat to the white stronghold. The South had lost the war, and now there were thousands of freed slaves everywhere, all of them trying to live the same way white people had thought was exclusively theirs. White Americans were enraged that in 1868 the Fourteenth Amendment granted black people the right to call themselves Americans, too, and that after the Fifteenth Amendment was ratified in 1870, black men could join white men at the voting polls.

Though Mack missed that league meeting, he vowed he would not miss another. But urged on by Jesse Billingsley, Governor Edmund Davis broke up the Negro Loyal League. Hostility between blacks and whites grew.

Jesse was a complex man. He had made sure his slaves learned to read, and he trusted Mack to negotiate his business dealings, but he seems not to have viewed black people as his equals. Though Jesse argued that Texas should not become a member of the Confederacy, he had not wanted to give up his slaves.

Throughout the 1870s and 1880s, black citizens stood up against the oppressive practices of embittered white citizens and garnered more and more power and influence, culminating in the elections of 1888. In Cedar Creek, the results were explosive: Two African Americans were elected to office. Orange Weeks was the new justice of the peace, Ike Wilson the new constable.

In June 1889, a few months after the new officials took office, a complaint was filed, likely by a black townsperson, against Frank Litton, son of one of Cedar Creek's leading white citizens. Ike Wilson arrived at Litton's house to deliver the summons, but Litton would not accept the papers unless a white officer served them. Wilson sent the deputy sheriff.

A few days later, when the trial convened in a residence near Cedar Creek Store, there were so many armed black men around the makeshift courthouse that the prosecuting attorney asked Justice of the Peace Weeks to postpone the trial. Weeks replied, "The white folks have had their day at running this court, and some of the rest of us will have ours now. The case will proceed."

Litton entered and sat down. His supporters stood behind him. All of the assembled, blacks and whites, clutched rifles and guns. The court clerk read the indictment and then Weeks asked the defendant to stand. Litton stood up, stated that he wanted a drink of water, and turned to walk outside (I think he was defying the black man's authority). Constable Wilson called out to him to halt. Litton kept walking. A gun was fired, and a bullet grazed Litton's head. Everyone began running and shooting. Weeks barricaded himself in a room; Wilson escaped through a window. Two whites and two blacks died at the scene.

Gramps was seven at the time. Excited and frightened by the sounds of chaos, he and a friend climbed to the top of the smokehouse. From their vantage point, they saw Mack scrambling and firing a pistol. They watched as he finally reached safety. When night fell and most of the gunfire had stopped, the children crept down from the rooftop and hurried home. As they burst through their doors, their parents grabbed and hugged them, relieved that the boys were unharmed.

White townsmen swore revenge on every Negro in Cedar Creek. Not only had they used their right to vote to elect former slaves into office, they also had firearms. Both ballot power and gun power were threats to the well-entrenched status quo. The sheriff ordered every citizen, regardless of color, to his home. Many black citizens fled town. Disobeying the sheriff, armed white men trawled the streets, broke into houses, and murdered black people at will. Cedar Creek did not calm down until the arrest and imprisonment of Orange Weeks and Ike Wilson in Austin. When news of the uprising spread, race riots broke out all over Texas. Several white people died. Many more black men, women, and children were murdered, though the numbers were not documented. Over the next few months, Billingsley, despite his displeasure about losing control over his former slaves, gave refuge to the Madison family.

Mack was literate, but it's not clear if he ever taught Martha to read. Both were likely too busy providing for and watching over their growing family. Between 1862 and 1884, my great-grandparents had ten children; only five — Charlie, Moody, Laura, Ruth, and John Chester, my grandfather — lived to adulthood.

The deaths, Mom told me, devastated Mack and Martha, but they had five children to bring up. And they had to stay strong in order to endure Jim Crow, which followed hard on the heels of Reconstruction. To keep their children and themselves uplifted, Mack and Martha repeated the family directive again and again. And they were free.

It seems that Charlie, the oldest, was not quite certain about how the saying could help him make his way through life as a black man in the South. He was the maverick. One evening, as he and a group of companions were leaving their jobs in the stockyard, several white men confronted him, accusing him of stealing a horse. A scuffle broke out, and one of the accusers died. Though no one knew who was responsible for the death, Charlie and his friends, fearing they would be jailed — or lynched — hurried out of town. That night, Charlie fled to Denison, about two hundred miles north. He hid in a friend's house for several weeks, changed his name to John Miller, and found work in railroad construction. White citizens harassed Mack and Martha for several months but finally gave up on learning Charlie's whereabouts. He kept in touch with his family but never went near Cedar Creek again.

While Charlie was adjusting to a new name and a new life, Moody found a steady job as a janitor at the State House in Austin. Laura, Ruth, and John Chester attended college — a rare accomplishment for black people in the early twentieth century. All three became teachers.

John Chester, my Gramps, the youngest of the surviving children, was born in 1882. Coreen's admonition echoed loud in his mind and heart until his death in 1960. As a teenager, Gramps dreamed of becoming a doctor. Mack offered him money for school, but Gramps didn't want to take money from his father, who had been a slave and worked hard to be able to buy his own piece of land, so he refused. Gramps had a plan — while picking cotton and doing odd jobs, he would earn a teaching degree, save every penny, then attend medical school. He did not foresee that jobs would be so limited and the pay so meager it would take nearly twenty years for him to earn his bachelor's degree in agriculture. He did not get to medical school.

After two years of teaching, Gramps became the principal of the Booker T. Washington School in Elgin, thirty miles north of Cedar Creek. It was the only school for black children in the town and surrounding area, and, as in many other southern schools of that era, there were only a handful of teachers for all twelve grades. During my grandfather's first year as principal, a young woman named Ruby Massey joined his teaching staff.

"Daddy and Mother met in a sad way," Mom began. I hadn't seen my grandparents often, and she wanted me to know them intimately. Gazing at the antique ring on her finger, she explained,

"He'd been in love with my mother's sister Ruby, and they planned to get married, but a few months after their engagement, she died."

The box held several photographs of Gramps taken decades before I had known him. In his youth, he was a man with erect posture, golden skin, wavy black hair, and a thick mustache. His face was serious . . . or so it seemed. Ruby, a small picture revealed, was a petite, full-bosomed woman with high cheekbones, brown skin, neatly groomed hair, and sparkling dark eyes.

I envisioned Gramps taking Ruby by the hand, walking with her, courting her, watching the way her eyes crinkled when he made her laugh. She accepted his marriage proposal and a ring. He likely foresaw Ruby holding and nursing their child and knew that any baby born of a love as strong as theirs would be a fine child. Then her appendix ruptured and took her life. Mom told me that, at first, he refused to believe she was gone. Then, finally, he grieved.

Years later, Gramps described his pain to his daughter and sons, and Mom always looked sad when she told me about it. "Daddy would stare out the window and say so softly I almost couldn't hear him, 'My clothes fell right off my body.'

"At Ruby's funeral," Mom would continue, "Daddy was so distraught that one of her sisters, Birdie Jo, went over to console him, even though they hadn't met before. At the time, she was living and teaching in another town. After the funeral, they got together on several occasions, mostly to reminisce about Ruby. Daddy was struggling to deal with her death, and Mother was someone who would listen to him and share his grief. They had two things in common — both loved to teach, and both loved Ruby. One year after the funeral, to everyone's surprise, including their own, most likely, they were married. Soon, they were teaching together in Elgin.

"So Ruby brought them together, even though they were two people who could scarcely be more different from each other. Daddy was full of hugs and kisses and couldn't resist any opportunity to be funny, especially when it came to playing around with words. Mother was standoffish and downright grim. Unfortunately for her, she was often the butt of Daddy's humor." Mother would smile. "Are you ready for one of your favorite stories?"

"The Ink Eraser!"

"Here it goes: One day when I was nine or ten, Mother sent Mack and John to Daddy's library to get some liquid ink eraser. The boys returned empty-handed. Twitching and stomping, they were trying so hard not to laugh I thought they would explode. Somehow, they managed to pass on Daddy's message: He had told them to ask their mother why she was so greedy. She looked perplexed. Mack and John explained, 'Look at us! You already have two *inky racers!*'

"They fell out on the parlor floor, rolling, giggling, clapping hands, slapping knees, clutching their bellies. Meanwhile, Daddy and I were hiding in the library, trying to muffle our laughter. Our cheeks were bursting, and tears rolled down our faces. When I peeked into the parlor, I saw Mother. She was livid. She had not gotten the ink eraser, and worse, far worse, the decorum of her home had been undermined yet again."

I never tired of this story, and it always made me laugh. Even more than Gramps's clever play on words, I loved knowing that his sense of humor had overturned Grandmuddy's somber rules. I wondered whether her rigidity and crankiness came from loving

Gramps and his family, circa 1935

a man who would forever love her deceased sister. And maybe his
humor was, in part, a way of getting by with a stand-in for the
woman he truly loved.

I had just turned nine when Mom took my two-year-old brother,
Biff, and me to visit my grandparents in Navasota. We had come
by train, but I knew the rules. We stayed in our compartment most
of the time and ate in the dining car after all the white passengers
had been served and departed. We did not eat behind a curtain,
and I did not see any girls to play with, only boys, whom I wanted
nothing to do with. The trip was really boring.

One afternoon during that visit, I was lying across my grandmother's bed, reading a book. My brother was taking a nap on
the living-room sofa. Grandmuddy's bedroom was behind the
front porch. She referred to the room as hers, and when I peeked
into the closets and drawers, I saw that they were filled with her
things only. There was no sign of Gramps there. I did not know

where he slept. Through the open window, I heard him talking to my uncle Mack. Only the backs of their shoulders and heads —Gramps in a frayed straw hat, Uncle Mack in a baseball cap— were visible to me. I could just make out a word here and there as Gramps confided in the younger of his two sons. He was talking about my grandmother. She was stingy, Gramps said. The sorrow in his voice told me that he was referring to more than her miserly ways with money. I cried into the nubby fabric of the bedspread. I knew my grandmother was humorless and rigid, and I felt sad that my beloved Gramps did not receive the love he deserved. Years later, after hearing the story about Gramps and Ruby many times, I came to realize that though Gramps probably learned to love my grandmother, Ruby never left his heart.

"Daddy hardly ever mentioned Ruby," Mom said, "but on my graduation day from college, he gave me this ring. All he said was 'Here's something I want you to have.' It was Ruby's engagement ring. I knew how much Mother and Daddy had loved Ruby. The ring and my name made me feel special."

17

New York Memorial

B efore returning to Montpelier, I went to Manhattan to see the African Burial Ground National Monument. The cemetery had been discovered in October 1991 when the excavation crew for a new federal building unearthed several human skeletons. A search through the city's archives revealed that those bones were the remains of a fraction of the thousands of Africans slaves who had lived there.

The visitors' center was a cramped, sparsely furnished office on the lower level of the World Trade Center. The staff was just beginning to define its relationship with New York City's black community and the people, like myself, who were pouring in from all over the country and around the world.

A pleasant full-figured young woman with dreadlocks hanging to her shoulders handed me a pamphlet. According to seventeenth- and eighteenth-century land surveys, the site had been part of a 6.6-acre slave cemetery. More than fifteen thousand bodies had been interred there. When planning the new building, the U.S. General Services Administration assumed that time, climate, and two hundred years of urban development since the cemetery closed in 1794 had destroyed the human remains.

In 1697, three centuries before the excavation, Trinity Church banned the burial of "Negro's [*sic*]" in its graveyard, thereby forcing the enslaved to find another place to inter their deceased. It was over this second site that the new federal building would be constructed. Early in 1992, when they learned of the discovery and the further damage to skeletal remains during the excavation, the city's African Americans formed the Descendant Community, mobilized, and chose that very church to be the venue for a televised meeting.

Adam Clayton Powell IV, a member of the New York City Council, stood at the microphone and told the U.S. government: "You do not disturb the deceased. You leave our people alone. You let them rest in peace. And if these reasonable and just demands are not met, then, at the very least, we should do everything that we can to stop the construction of the building."

The Reverend Dr. Herbert Daughtry reminded the nation that, "had it not been for the bodies and the bone, the body and the labor, of those people who rest yonder — our ancestors — there would not have been a United States of America. There would not have been any wealth in the Western world. And not only that, after we had worked the fields and built the roads, and had our bodies sold, we went out and died for the country."

I was particularly stirred by the proclamation of city council member Helen Marshall: "We were the slaves, and our magnificence is in our survival. What is in that burial ground will teach us immensely." Mandy and her enslaved descendants had survived, and someday soon, I would return to Montpelier and find her grave. She had much to teach me.

Newspapers and television newscasts throughout the world reported the night vigils, the organized rallies, the spontaneous demonstrations, and the petitions and meetings. Many of the country's black citizens were enraged that their government would construct a building on top of their ancestors' final resting place. Their people had been devalued and enslaved in a nation founded on the claim that "all men ... are endowed by their Creator with certain unalienable Rights that among these are Life, Liberty and the pursuit of Happiness." Spiritual leaders of many faiths came to the sacred site, and the cause became a cry for human rights worldwide.

One year after the discovery, in October 1992, Congress passed a law to alter the planned building's design in order to preserve the archaeological site. The legislative body then appropriated three million dollars for a museum and research center. That year, the African Burial Ground was listed on the National Register of Historic Places, and in April 1993, it became a National Historic Landmark. The graveyard is the largest American colonial cemetery for enslaved Africans and perhaps the earliest and largest cemetery for any ethnic group.

Yielding to unrelenting protests from an alliance of scholars, politicians, the Advisory Council on Historic Preservation, and the Descendant Community, the House Subcommittee on Public Works agreed to transfer the 419 exhumed skeletons to Howard University in Washington, DC. Under the direction of Dr. Michael Blakey, an African-American physical anthropologist, the bones were separated, labeled, cataloged, and packed. In a nighttime ceremony at the burial site, with hundreds of participants and onlookers, Dr. Blakey, in traditional formal African attire, accepted a small box wrapped in African fabric that contained the last of the remains to be transferred.

At Howard University's Montague Cobb Biological Anthropology Laboratory, Dr. Blakey and his team analyzed the remains and learned the nature and extent of the brutality suffered by the black men, women, and children who had come to New York from Angola, the Congo, and the Caribbean. The researchers studied the bones, teeth, and hair of the buried individuals and discovered that almost half were children under twelve years old and that more than half of these had not reached their second birthday. The children, like the adults, had suffered from malnutrition, injury, infectious diseases, lead poisoning, and overwork.

But the slaves had buried their loved ones in accordance with centuries-old African traditions and beliefs. Wrapped in winding sheets secured with shroud pins, laid supine in individual coffins, and buried with the head at the west end of the grave, the deceased could now achieve the status of ancestor.

Some skulls had coins in the eye sockets. Among the bones lay many artifacts, including buttons, bracelets, cufflinks, and lots of beads. The dead had been laid to rest as human beings.

A boy under the age of five lay with a clamshell above his left collarbone. Perhaps this child had loved the ocean. In another coffin, a woman lay with her newborn infant nestled in the bend of her elbow. When I saw a photograph of their skeletons, I imagined the dying mother struggling to nurse her dying child, and I could feel the love and pain surrounding them as they lost their battle against death.

During my first visit to Montpelier, I'd forfeited the chance to visit the slave cemetery. Now, learning about the burial practices of New York City's enslaved people, I resolved to return to Virginia.

18

The Plantation's Tale

I n April 1997, five years after my first visit to Montpelier, I returned to find Mandy's final resting place, the end of her journey through the earthly world. I stopped off in the town of Orange, strolled the gently sloping side streets, and enjoyed the charm of small colonial houses. Most of the homes in the town of some four thousand residents were white clapboard with black mullions and doors. Many had window boxes or hanging baskets newly planted with colorful arrays of spring flowers. The limbs of budding oaks draped over sidewalks and lawns.

I drove past the county courthouse, built in 1859, and stopped at the Orange County Historical Society to see Ann Miller. I found her sitting at a large oak table in the center of a stuffy room lined with bookshelves.

"Good to see you again," Ann said the moment she looked up. "It's been way too long."

"I've been traveling — Portugal, Ghana, Maryland, New York, Texas."

"And back to Virginia."

"Yes. I didn't quite finish what I started five years back."

"Well, I've been right here, doing a little research consultation for the historical society and working full-time for the Virginia Transportation Research Council, better known as VTRC. I try to keep busy, but I miss T.O. He died a few months ago," Ann said. "Natural causes."

"I'm sorry to hear about T.O. He reminded me of my grandfather, and as I was leaving, he gave me a hug. I regret that I couldn't come back for his ninetieth birthday party."

"It was quite a celebration. He invited everyone he knew and everyone he met," Ann said. "We all miss him. He was a great guy, and he and his fascinating personal story helped keep Orange County history alive."

"That trunk he found makes it so tangible."

"His family is looking for a place to display it — maybe a museum in DC."

"I parked across the street, right in front of the James Madison Museum. What's in there?" I asked.

"Let's walk over."

"Welcome," the silver-haired, bespectacled receptionist said, reaching across the counter to shake my hand. "Walk around, and take your time. You don't want to miss anything."

"Start with the door at the back," Ann suggested. "Open it. You're in for a surprise."

Ann returned to the historical society, and I walked to the rear of the museum and opened a wooden door. Suddenly, I was a time traveler who had landed on a platform overlooking a cavernous room. A full-size, two-story house stood below me. I descended the steps leading to the floor of the Hall of Transportation and Agriculture.

According to a sign in front of the dwelling, the sixteen-cubic-foot "cube house" was big by the standards of the time. A family of eight, or even more, would have considered themselves quite comfortable there. The Arjalon Price House, built circa 1733, had one room with a fireplace on the first floor and a narrow winding stairway leading to the one room on the second floor. There was no kitchen or bathroom in the house. The women prepared food in a separate building; during the eighteenth century, bathing was not a common practice, and in most households, chamber pots were the only toilet facilities.

Several aging wood and leather horse-drawn carriages stood around the cube house. Massive plows and crude hand-carved tools were scattered across the floor of the huge, barn-like room. These vestiges of a bygone era looked more prehistoric than preindustrial to me, their use time- and labor-intensive. I imagined mules and slaves forcing the plows through Virginia's red soil, leaving rows of furrows and mounds in their wake.

After nearly an hour looking at the relics on the floor of the vast hall, I ascended the stairs back to the main section of the museum and wandered the tight, low-ceilinged rooms. Framed documents and photographs of nineteenth-century Orange County citizens and Civil War battalions lined the walls. Artifacts once belonging to James and Dolley Madison — china, crystal, silverware, lace collars, and books — were displayed in glass cases, unlike the treasured photographs, letters, and documents in the box my mother had brought for me to hold in my hands, learn from, and pass on. I could not touch James and Dolley's items, but they, too, were among my family heirlooms.

I was about to leave the museum when a notice captured my attention:

> To be Sold, on Tuesday the 15th of May, at Orange Court-House
>
> EIGHTY likely Virginia born SLAVES
>
> Consisting of Men, Women, and Children, among whom are a great variety of likely young wenches and fellows, for ready money, bills of Exchange, or good merchant notes, payable at the next Court of Oyer. There are sundry carpenters, a good blacksmith, and a master collier. A satisfactory title will be made, and the public may depend the above number (at least) will be produced and sold, let the weather happen as it will, by Joseph Hawkins

I felt the same tightness in my chest I'd experienced on my first visit to Montpelier. I could not comprehend how the skill and talent of "carpenters, a good blacksmith, and a master collier" had been valued without any consideration of the humanity of the men who had these abilities. The tightness became a crushing ache, and my hands quivered as I took in the fact that children had, "for ready money," been "produced and sold." Some of the "likely young wenches" may have been prepubescent, unaware of or, for some, fearful of the risks inherent in their impending black womanhood.

There was no date on the notice, but the words *likely Virginia born slaves* told me that Mandy had not been among the eighty. Yet while envisioning her alone and scared, waiting to be called, I began to imagine myself standing on the auction block, naked except for the stained white cloth around my hips, the bids deciding my worth.

I hurried out of the museum to my car and sat for several minutes, taking slow, deep breaths, before I started the engine. Finally, when I could breathe more comfortably and my hands shook less, I eased the car away from the curb and headed toward Montpelier. Along the way, I stopped from time to time to calm myself in the peacefulness of picturesque farms, neat cottages, and a dense pine forest whose tree branches formed ornate patterns of shadow and light that flickered across the road. At the end of the thirty-minute journey, feeling grounded in the twentieth century once more, I found a shady spot to park near the visitors' center, then took the shuttle bus up to the Madison plantation. Once again, I felt I belonged there, as if I had come home.

The president's family had a long history in the area. In the spring of 1732, Madison's grandparents Ambrose and Frances brought their son and two daughters to Orange County, a region of lush, rolling countryside thirty miles southeast of the Blue Ridge Mountains. Ambrose had sent his slaves ahead to clear the land for a tobacco plantation. Mount Pleasant included a two-room house, a freestanding cookhouse, barns, storage huts, and slave quarters.

Within a few months of the move, Ambrose was dead — poisoned, it was said, by Pompey, a slave from a neighboring plantation, assisted by Dido and Turk, two of Ambrose's own slaves. This was the first documented murder of a master by slaves in Virginia. Presumably Dido, a woman, and Turk, a man, agreed to help Pompey because they were upset with how Ambrose treated them, and it is likely that the "cure" for their mistreatment was a centuries-old West African potion.

Killing a master was an act of treason, punishable by death. The accused proclaimed their innocence, but the court found them guilty. Pompey was sentenced to death by hanging. Dido's and

Turk's only punishment was twenty-nine lashes each, in part because Ambrose's widow, Frances, needed able hands to help farm the approximately five thousand acres now in her charge.

Twelve years later, at the age of twenty-one, James, Frances and Ambrose's son, took over the property and bought more land. Despite having received little formal education, James Madison became a successful businessman and "gentleman farmer."

The year 1749 witnessed the union of two prominent, well-established Virginian families: the Madisons and the Conways. James turned twenty-six that year. His bride, Nelly, was seventeen. In 1751, Nelly gave birth to James Jr., the first of the illustrious couple's twelve offspring, only seven of whom would survive to adulthood.

James Madison Sr.

In the mid-1760s, James Sr. built a new house near Mount Pleasant. As he obtained more land, money, prestige, and power, he also acquired some one hundred slaves. And the personal papers of George Fraser (the original holder of Sarah Madden's indenture) reveal that Fraser, Madison, and other area merchants had formed a partnership to buy and sell slaves locally and regionally. This business arrangement was referred to as "the Negro Concern."

The family's size and prestige grew, and the house that James Sr. started was enlarged again and again to accommodate grander and grander needs and desires, symbolized by the four towering columns that now adorn the sweeping front portico, added by James Jr. and his wife, Dolley, in the late 1790s.

At the time of my visit, the home, once a stately testimonial to the wealth generated from the thousands of fertile acres that had surrounded it, was now a memorial to an American president, to his ambitious family, to the slaves who built it, and to the southern way of life that was destroyed by the Civil War but never forgotten.

During most of the eighteenth and early nineteenth centuries, Virginia's wealth came from tobacco. In the latter half of the eighteenth century, tobacco prices fluctuated rapidly and wildly. To secure financial stability, James Madison Sr. ventured into ironworks, building contracting, and distilling his renowned peach brandy. He also diversified Montpelier's crops to include hemp fiber and wheat.

Three decades after Madison Sr.'s death, in 1801, a general depression of tobacco prices set in and lasted throughout the 1830s. Within a few years of Madison Jr.'s death, in 1836, prices had reached bottom. Dolley had no reliable source of income, and her son, Todd, had squandered much of what remained of the Madison fortune. In 1844, to ease her strained finances, Dolley sold the plantation.

Over the years, a series of owners removed walls, altered windows and doors, and installed plumbing and electricity. At the turn of the twentieth century, members of the duPont family, the chemical-products dynasty, purchased Montpelier. They used the plantation to raise and show horses, adding stables and racetracks around the main house. The duPonts also added a second story to the Madisons' one-story wings and converted a large room in the house into a showcase for their trophies and photographs of horses, riders, and trainers.

In 1983, Marion duPont Scott died; she left the mansion and its twenty-seven hundred remaining acres to the National Trust for Historic Preservation. Inside the mansion and behind the coral-colored stucco that covered the original red-brick façade, architects and historians probed for secrets. Like the slaves who had built it, the house had endured the ambitions of owners, its value to American history often obscured. Among bricks, behind plaster, amid tobacco roots, buried under manicured lawns, and in unmarked graves were the stories of Montpelier's enslaved people.

The resident archaeologist Lynne Lewis was at an excavation site on the former plantation when I arrived, so I waited for her on the front portico. Though the April weather in the foothills of Virginia's Blue Ridge Mountains had been erratic, on the day of my return, the sky was clear and blue, and the sun was just beginning to warm the morning air. The distant, densely forested mountains that spread out before me were blue, almost purple, and so wide they wrapped around the horizon. Spring here, I thought, was alluring with its promise of warmth and sense of awakening. Summer, with broad fields scattered with a colorful collage of wildflowers stretching from the mansion toward the hills, had not yet arrived, but already I imagined autumn's soaring trees ablaze with red, orange, and gold leaves. I foresaw rolling outlines of silver-white winter snow rising gently along hillsides toward the sky's changing hue. Beyond the crest, snow glided down the slopes into the black lace of naked dormant trees. More than two centuries earlier, Mandy, Coreen, and Jim had witnessed the seasonal drama here.

Despite its ever-changing beauty, Montpelier had been a prison. The landscape around the mansion was sweeping and open, but I felt the confinement of the plantation's antebellum years. Once, scores of slave cabins lay south of the mansion, and I pictured them scattered low on the hillside, leaning perilously, their weathered roofs slowly giving way to the pull of the dusty earth. The sinking roofs mimicked the bowing hills, all succumbing to the forces of the land. Meanwhile, slaves, diminutive in the distant fields, toiled, their labor coaxing riches from the soil, making possible the Madison legacy.

I went to the back of the mansion, hoping to step again into the furrow where Coreen had walked, where, together, she and I followed a visible, dusty red path into landscape that had held black people hostage to ambition and greed. The path was gone. A brick walkway now covered Coreen's footsteps, burying my physical connection to my enslaved ancestor. I felt alone and unmoored.

Lynne found me wandering aimlessly. "Sorry to keep you waiting," she said.

"No problem," I murmured. I did not mention Coreen's buried footsteps.

Silently, we headed toward the slave cemetery. Scattered pebbles on the dirt road glistened in the morning sunlight, and wide meadows speckled with young daffodils sloped green and bold. But as Lynne and I walked the thirteen hundred feet from the mansion, I grew more and more afraid I would see an isolated patch of dry soil with clusters of poison ivy and dandelions, a place where in death, as in life, slaves had been dishonored. Bursts of wind made the trees shudder. I huddled into my jacket.

"I'm sure that at least one of my ancestors lies buried here," I said, trying to reconnect to them.

"Do you know much about them?"

"Very little, but one was called Mandy. No one knows her African name. She was stolen from Ghana and ended up in a cotton field in a remote part of this plantation."

"But Virginia is a tobacco state. If there was a cotton field here, there's no sign of it nowadays," Lynne said in her raspy voice as a breeze swept her short blond hair off her forehead.

"Mom told me it was small and that the cotton wasn't very good, so it was only used to make 'Negro cloth' for the slaves to make their own clothes."

"Sounds possible. We know a lot about Montpelier, but we certainly don't know everything," Lynne said as we rounded a curve. "We don't even know who's buried here. Might be Confederate soldiers."

I stopped short. Lynne turned to face me. "Confederate graves are scattered all over the South. Men died in battle and were buried near where they fell," she explained.

"This might not be a slave cemetery?"

"I think it is, but some folks around here disagree." We walked toward a low hill that rose behind a cluster of trees. A moment later, Lynne said, "This is it."

The field of bright blue periwinkles took me by surprise.

We stepped off the dirt road and entered a small woods. Blanketed with fallen leaves and cradling the hidden remains, the ground was soft underfoot.

"Because of these periwinkles," Lynne told me, gesturing toward the flowers poking up through the leaves, "we know this is a burial ground. Back then, people in the South planted them to beautify cemeteries. Under the leaves, we found linear depressions where the graves had sunk a little, all of them oriented east to west. Then last winter, there was an early thaw. The only snow left was in the depressions. Eerie and beautiful white ribbons."

Lynne guided me farther in, and we were careful not to step on the graves or kick the headstones and footstones, crude blocks of white quartz. One was streaked with brownish-orange veins and mottled with red pits. It looked like dried blood. I sensed my ancestors trying to tell me something.

Lynne and I stood in a kingdom of many trees, but only one of them called to me. I approached it, reverent. The trunk was broad, the thick branches pregnant with pale leaf buds. I touched the rough bark. Nestled at the base was a stone the color of raw flesh, its uneven surface shiny and smooth.

I knelt down and touched it. My hand trembled, and I envisioned my family's first *griotte* lying beneath, draped in white muslin. In her callused hands rested her sole memento from her homeland: a single red bead.

The gentle morning sun was warm on my back, countering the chill in the air. My shadow fell over the rock, uniting me with my family's first matriarch, my five-greats-grandmother. After several minutes, I asked, "Were the bodies placed in coffins or wrapped in winding sheets?"

Lynne replied, "We don't know, and we never will. We would

never do anything to disturb the people lying here." Her statement embraced the sense of mystery and peace in our surroundings, and I realized that to find the truth, someone would have to disturb this place of final rest.

"The headstones are at the western end of each grave," Lynne said, "so the spirits of the dead can rise facing east when the Savior returns in the Second Coming."

"But that's a Christian belief."

"Yes, but see that curved, low earthen ridge around us on three sides?" When we reached the center of the cemetery, she continued. "Only part of the border is the natural lift of the land. We've studied African burial traditions, so we think slaves built up the rest of the berm to create a haven for their deceased loved ones."

Lynne explained that in order to keep demons away, the slaves had made sure there were no right angles in the ridge safeguarding the graves. Within this rounded sanctum, the souls of ancestors dwell, taking part in the everyday tragedies and celebrations of the living and providing protection from evil spirits. The east–west orientation of each grave is in harmony with the African belief that the patterns of life and death follow the movements of the sun. Mankind awakens as the sun rises in the east, then he works, struggles, dances, plays, sings, cries, laughs, and loves until the sun has set in the west. The placement of the graves might also suggest that the slaves had woven Christian doctrine into traditional African customs.

The slaves chose a wooded area because, through the generations, the elders and the *griots* had taught them that the roots of trees guide the spirits of the dead to the realm of the underworld. To prepare the soul for its journey, mourners placed the last item the deceased had touched onto the body. At midnight (midday in the land of the dead), the departed's strengths and talents leave the object and enter the dreams of descendants in order to inspire them.

My mother's stories had set me on the path that led to this very spot. Mandy's hands had sown and reaped the riches of southern soil for someone else. Mandy's womb had given birth to descendants who had harvested her strength to become farmers, carpenters, teachers, police officers, civic leaders, entrepreneurs, lawyers, dentists, doctors, nurses, social workers, engineers, psychologists, ministers, railroad porters, salespeople, musicians, and artists. "We have wings," Gramps had told me.

In the seven years since I'd received the box, I had struggled to become the *griotte* my family deserved and to understand why, or even whether, I should feel proud to be a Madison. The quest, started by my mother, would never be over; there was much more to find out. But I had become the *griotte*. When the time came, I would pass the box and the quest on to my daughter, Nicole, along with what I had learned: Death is inevitable, but as long as we who are left behind remember our ancestors, they are part of us and remain alive to us and teach us.

The story of my African-American family is ten generations long, and there are many people in our two-hundred-and-fifty-year saga, including villains and heroes, the famous and the obscure, the powerful and the powerless. But it is Mandy who is my inspiration.

Mingling with the sounds of the graveyard and my own breaths, Mandy's voice rose from my thoughts. I looked up and said with confidence to Lynne, "Mandy lies here."

Mandy

Mandy was not my name. Never.

My mother often told me, "You were born screaming, ready to fight."

She said that as I was arriving in this world, my father walked back and forth under a blue sky and a bright sun.

"No men allowed!" the midwife shouted at him whenever he got too close to the opening of our hut. Finally, after many hours, she came out and grabbed his hands.

"A girl!" she said.

He took one step inside but realized he had forgotten the defomo dan, the hand-washing rum that was also the gift for everyone assisting my birth, and he had forgotten one of his cloths to put under my head. My father couldn't come in without the cloth because it would mean he didn't want to recognize me as his child. And he did!

He ran back to the tree where he had left everything, then rushed back to the hut, a gourd of rum in one hand and the cloth in the other. Again, he took one step inside, then another and another. With each step, I screamed louder, my mother said, but when he folded the cloth and placed it under my head, I stopped crying. He smiled. I blinked and then fell asleep.

As was our tradition, I was kept in the hut for eight days, carefully watched, to see if I could survive the many dangers outside. Finally, it was time for my outdooring ceremony, the proclamation and naming, my kpodsiemo. *Until then, I was unknown, a stranger to the world.*

At two o'clock in the morning, my mother wrapped me up, and a pair of elderly women from my father's hut carried me, the moon lighting the way, to my father's father. The entire village was there in front of his hut. My grandfather took me in his arms and unwrapped me. I whimpered. Then he held me up and brought me down gently to the ground and sprinkled me with water three times. My father climbed on top of the hut and poured a calabash of water on me so that I would know rain and the earth. I screamed, and everyone laughed, he said. I didn't stop crying when my grandfather, the oldest man in the village, prayed:

> *May the gods pour their blessing upon us! May the gods pour their blessing upon us! May the gods pour their blessing upon us! A child has been born; we have formed a circle round to view her. Whenever we dig, may it become a well full of water, and when we drink out of the well, may it be a means of health and strength to us! May the parents of this child live long! May she ever look at the place from whence she came! May she be pleased always to dwell with us! May she have respect for the aged! May she be obedient to elders and do what is right and proper. May many more follow, full of grace and honor! May the families always be in a position to pay respect and regard to this child, and out of her earnings may we have something to live upon! May she live long and others come*

*and meet her! As a Ga person does not speak at random, so
may this child be careful of her words and speech and speak
the truth so that she may not get into trouble May the gods
pour their blessing upon us!*

I received many gifts, among them shiny red beads from my
mother. Then my grandfather announced my name to the whole village. He had chosen it weeks before my birth. He had thought and
thought, my father told me, to find something to make him and my
family and the village proud. He'd consulted our ancestors so that my
name would bring pride to them too.

My name is a story. There is a message in each part, but the power
is in the whole. My name tells the day of the week on which I was
born, my soul name. My name represents my tribe, my father, and
my birth order. My name tells the history of the family that showed
me how to be who I am and to value family above all else. I also have
a name for the special qualities my grandfather chose for me. This is
my den pa. He chose hopefulness and inner balance and strength for
me. I have a pet name, too, one that only my parents and grandparents use. My love name.

My name is my secret. I keep it hidden inside me in the sacred,
immaterial vessel I was born with. It holds the spirit of my ancestors,
the blessings of my village, and the love of my family. It holds my
hopes and my dreams for my children and my children's children.
My secret connects me to my ancestors and my parents, to Coreen
and Jim, to Emanuel and Mack, to John Chester and Ruby, to you,
Bettye, and to Nicole and her babies, Peter Lee and little Madison.

Secrets are power to those who know the secret. I was stolen, and
I never told anyone the name my family had blessed me with at my
kpodsiemo. I did not bring damnation to myself, my family, and

my ancestors. But when I lay on the bottom of the boat that took me far away, stood almost naked before men who wanted to own me, labored up and down rows of cotton and tobacco, or lay under the terrible weight of the man who violated my womanhood, my kpod-siemo *was the only name I called myself.*

Massa called me Mandy.

Massa called me slave.

He was wrong.

I am griotte, *master of eloquence, the vessel of speech, the memory of mankind. I speak no untruths. This is the word of my father and my father's father. Listen to me, those who want to know. From my mouth you will hear the history of your ancestors.*

19

History, Heritage, Memory

By 2013, I had spent more than twenty years traveling, researching, and writing about my ancestors. Whenever I thought about giving up, I relived the moments I walked in the furrow behind Madison's mansion, tracing Coreen's footsteps with my own. She was there with me; I could not let her down. I was excited every time I held my copy of Emanuel and Elizabeth's "marriage certificate." I could feel the awkwardness of their first meeting and then their love for each other and their children that transcended the hardships of enslavement and the failures of Reconstruction. I had followed Mandy's path through the world. The losses and the abuses, including rape, she had experienced would have shattered a lesser person. Hers was a story of strength that demanded to be told.

But when I tried to find proof that she and Coreen and Jim had existed, I ran into one roadblock after another. Without DNA data or archival information to verify the stories told in my family for more than two hundred and fifty years, how could I get anyone to believe me? How could I get anyone besides my own family to care?

In March 2014, I accepted an invitation to James Madison's former residence to participate in a workshop, Interpreting the African-American Landscape at Montpelier. Since my initial visit in 1992, I had returned several times, developing an emotional connection to the plantation. This was where my black ancestors, held in lifelong bondage, had lived, worked, and contributed mightily to the success, wealth, and prestige of the Madisons and of America. This was where my slave-owning forebear James Madison Jr., the man who became the Father of the Constitution, the secretary of state, and the fourth president of the United States, had read volumes, formulated thoughts, and written reams.

In 2001, at the first "re-membering" of Montpelier slave descendants, I shared my family's oral history and the search for evidence of my enslaved ancestors, and in 2007, I helped organize the second gathering, so at the 2014 workshop, as I joined other Montpelier slave descendants, the local African-American community, and college educators, I felt at home.

The purpose of the two-and-a-half-day retreat was to help the leadership and staff of the historical site to, in their words, "better interpret the physical landscape — the grounds, slave quarters, kitchens, freedmen's homes, and other dependencies and farm buildings — so that we can more accurately reflect the lives of the African Americans who lived and worked here in the era of slavery, through Reconstruction and Segregation, and into the modern era." We toured the estate, listened to speakers, participated in brainstorming sessions, and served as advisers.

The focus was to bring visibility to what recorded history had obscured. We would uncover and promote the importance of recognizing, accepting, and valuing a different kind of national narrative. We honored the vital contributions slaves had made to the United States and its dominance throughout the world. We spoke about seeing beyond the story of our country as it had been conceived and written from the select perspective of select people. Their elite voices, beliefs, and values, legitimized by the records they created and kept, shaped what our nation considered its history. But what of the stories lived and told by people whose lives and thoughts were excluded from the dominant narrative that molded America? Those stories, kept alive by *griots* and in personal tales and family memorabilia, are also our nation's stories and should be brought to the forefront and shared.

In the workshop's opening presentation, Christy Coleman, director of the American Civil War Center, wrote three words in bold letters across the whiteboard: HISTORY, HERITAGE, MEMORY. An attractive brown-skinned woman and the only one in the room wearing high-heeled shoes and a business suit, she was a former actress who knew how to engage her audience.

"*History*," she said, tapping the word on the board, "is what actually happened. *Heritage* is what a community has chosen to embrace and what tells that community why it should care. *Memory*," she said as she walked to that side of the board, "is the intimate, individual connection to the story. It is the oral narrative in your immediate circle, your family.

"Imagine," Coleman said as she looked at each of us, "James Madison carefully crafting the language of our Constitution, crafting it to be a living, breathing document, being very careful to never use the word *slave* or *slavery* but rather *other persons*."

There was a collective intake of breath, and I watched Coleman watch us as we absorbed this fact. How had I not paid attention to this? I wondered. How was it possible that in all of my American history classes, no one had ever discussed what the authors had done to people of African descent when they excluded two critical words from the document that defined the nation's image? Their omission made it possible for white Americans to take a figurative look into a mirror and like what they saw and for people around the world to admire a just, compassionate, civilized new democracy. But their omission, I knew, did not mean that slaves and slavery were not vital to America's existence.

Christy Coleman wasn't finished. *"Other persons,"* she said, "is an extraordinary choice of words, because what does it do?" She stepped toward us. "It contains a humanity that was going to have to be dealt with."

That evening, I returned to my room, questioning the suggestion that the wording of the Constitution acknowledged that slaves were human beings. I found seven items dealing with slaves and slavery, beginning with article 1, section 2, the one Coleman had referenced:

> Representatives and direct Taxes shall be apportioned among the several States which may be included within this Union, according to their respective Numbers, which shall be determined by adding to the whole Number of free Persons, including those bound to Service for a Term of Years, and excluding Indians not taxed, three-fifths of all other Persons.

This precedent-setting instrument of government made concrete the perception that African slaves were lesser. By talking around slavery and using the word *persons,* the Founding Fathers had inadvertently, and perhaps counter to their intent, admitted that the nation's enslaved were human.

And yet, this choice of words rendered slaves invisible, a vague "other," no more than an economic necessity and a political tool unworthy of the justice and blessings of liberty the Constitution promised the people of the United States. The real story was not about from whom African Americans descended. The real story was that our stories had been left out. More than ever, I understood that this omission was why oral history was essential to African Americans having knowledge of how crucial we have always been to what this nation is.

Coleman's articulation of the differences among history, heritage, and memory illuminated for me that I was not alone in my efforts to bring visibility to the invisible, that there were people outside my family who would care about our story. Her discussion of "other persons" helped me realize that the problem I faced in finding my enslaved ancestors was not DNA; the problem was the Constitution.

In his keynote address the following evening, Rex Ellis, associate director for Curatorial Affairs at the Smithsonian's National Museum of African American History and Culture, looked and sounded like the southern black preacher he in fact was. Ellis described James Madison's world as one that depended on "a community of people who surrounded him, a community of people who supported him, and a community of people who challenged him and reminded him [that] they, too, were contributors to all he was able to accomplish.

"When he woke up in the morning," Ellis told us, "they were the ones who summoned him. When he washed his face, they were the ones who brought the water. When he ate, they were the ones who prepared his food. When he was sick, they were the ones who cared for him and nursed him back to good health. When he went to the White House, they went too ... They were the ones who were by his side when he died. They were the ones who challenged him as a slave owner, who yearned for something more than what they had, the ones who reminded him, in their own ways, that their bodies might have been bent, but they still desired to be free."

Ellis told us that the work to which we were contributing acknowledged that "history is messy; history is frustrating; history is oppressive; it is imprecise but eminently worthy of our best pursuit ... We are all part of the historical process. It does not end at the death of any one man or any one woman. It continues, and we are the ones responsible for its legacy." Paraphrasing Bernice Reagon, Ellis concluded, "We are the ones we have been waiting for."

It felt good to be part of the historical process, but finding and proving the truth is often daunting. In Lagos, Portugal, a concession stand mocks the suffering that took place there. Aboard slave ships, captives had no names. In New York City, evidence of slavery in a northern city rests beneath the Ted Weiss Federal Building. At Montpelier, Coreen's footsteps lie under a brick walkway. Wherever I turned, missing, burned, hidden, or nonexistent records led me to dead ends.

Across the generations, our family *griots* have had vision, hope, and perseverance. We have told and retold our stories, and, through whatever means available, we have taken on the responsibility of affirming the existence and the accomplishments of the

Other Madisons. Mandy's story would have died with her if she had not learned that staying alive meant being "fighting mad." Coreen, by words alone, taught her son that in order for there to be any chance for them to see each other again, he would have to remember his name was Madison. Her son and grandson, Jim and Emanuel, were literate, but because slave owners did not believe that their slaves should learn to read and write, my enslaved ancestors relied on the ancient West African tradition of oral history. If these two literate *griots* had come across written records, there would have been no place to keep them. Even their pockets were not their own.

Nine years after emancipation, my great-grandfather Mack, Jim's grandson, held an important document in his hands: the title to his own land. He saved the deed in his Bible. Mack's desire was for Coreen's entreaty to inspire future generations. In his old age, as he handed over the Bible and, with it, the duties of *griot* to his youngest son, my Gramps, Mack repeated his iteration of the credo: "Always remember — you're a Madison. You come from a president."

Once a month, the colored citizens of Elgin, Texas, could use the public library. Gramps was there when the doors opened and stayed until dark, looking for information about West Africa, the slave trade, and slavery in his country. He did not look for his own ancestors in library books; he knew they were not there. Gramps was of the first generation to be born free, and the first time he told his children the stories about his enslaved ancestors was when he changed the saying to "Always remember — you're a Madison. You come from African slaves and a president."

Though my mother never touched a computer, she witnessed

the ways the electronic age changed how people communicated and whose voices were heard. She made her intention clear: "I want to give you plenty of time to write the book."

I believe my mother recognized that the opportunity for our family's story to take its place in recorded history had finally arrived. I also believe she knew that as the hardships endured by our family during slavery, Reconstruction, and Jim Crow receded from our minds, the risk increased that our stories would be lost and our ancestors forgotten.

I could not let anyone hide my footsteps. To honor Mandy and Coreen and all the Other Madisons, I had to put aside my reserve and join them in the emotional journey that made us who we are. Living in a time of rapidly changing technology that allowed diverse voices to speak out, I had an unprecedented opportunity to contribute to the national historical narrative. With my mother's words urging me on, I would make some noise so that my ancestors would be heard and remembered.

As I researched my family and traveled to places where horrific things had happened, friends told me that what I was doing to reveal the truth about the Other Madisons was courageous. It did not feel like courage. I was looking for Mandy. After walking in Coreen's footsteps and realizing how connected I could feel to my ancestors, I wanted to walk where Mandy had walked and see what she had seen. I tried my best to grasp what she went through and what she thought and felt. I assumed knowing her deeply would help me become an unflinching *griotte* who understood and had reconciled all that it means to be among the Other Madisons. In the end, however, I will never know what it was like to be stolen and put on an auction block, to lose everything I knew and everyone I loved, to be vulnerable to someone else's power, to be a slave for the rest of my life. If I could have been Mandy for ten minutes, or however long I could bear it, I would have done it. I hoped that I could convey her strength, do justice to her life, and keep her memory alive. I hoped to find the words to make the world pay attention to her.

This book will be my legacy, my life's purpose, and I share it with Mandy and all who came after her. I share it with other slaves and their descendants. I offer it to all Americans who care about the gross mistreatment of millions of African slaves and who believe, as Gramps believed, that African Americans have always been vital to this nation. In Gramps's words:

> Our white ancestors laid the foundations for this country, but our dark-skinned ancestors built it. They worked the fields, nursed babies, preached sermons, and fought in wars. They played music, owned businesses, cured sickness, and worked on railroads. They taught their children everything important about life in this world. They taught their children about God.

The story my family kept alive and those that other African-American families kept alive have evolved from memory to heritage and now can emerge as history, a more inclusive and complete history. In order to make the transition, we cannot allow our footsteps to be buried under someone else's disregard or be wiped away by tides of ignorance. We cannot allow our presence to be obscured by racist complicity, our women disrespected, our young people discarded. And, for our own part, we cannot allow ourselves to shrink away from the painful facts in our family histories.

Our magnificence is indeed in our survival, but it's also in our intelligence and creativity, in our self-respect and self-determination, in the potential of our young people, in our importance to America's global success and global presence. It is in our stories. President Madison's black descendants, his only descendants, are not "*other* Madisons." And America's slave descendants are not "*other* Persons." This message is the cornerstone to what I bring to my family's oral tradition.

I have been asked many times, "Would it matter to you if you learned, with no uncertainty, that you are *not* a descendant of President Madison?"

My friends know how long and hard I have tried to find Madison's son, Jim. They have shared my disappointments and frustrations. My friend Renée, who is white, said, "You *are* a Madison. Why do you have to prove it?"

I could have answered, *Because I'm black. I doubt the recognized descendants of Madison's family get the same question, and I would be shocked if anyone asked them, "Would it matter to you if you learned, with no uncertainty, that President Madison fathered a child with a slave woman?"* Instead, I smiled and said to Renée, "You've always been in my corner. That's why I love you."

However, when casual acquaintances ask me whether it would matter to me if it turned out I have no connection to the president, I detect their assumption that my family story cannot be true and, moreover, that we are seeking a way to rise above the presumed limitations of being black in America. Nonetheless, I answer the question.

"No," I reply, "it would not matter. I've always known I am a Madison, but when my mother handed me the box of family memorabilia, the journey of becoming the *griotte* changed me.

"The Madison name," I explain, "was important to my family's early story. My sold-apart family members hoped to use it to find each other, but they died before they got the chance. Then, when freedom came, the name inspired my enslaved family to do great things, just like their famous ancestor. But my grandfather, who was born free, admired his parents and his aunts and uncles and his grandparents. They survived slavery, he told his children, because they were strong inside and believed in themselves. 'And,' he always added, 'you're just like they were. Strong. Smart. Capable of doing anything you set your mind to. Nothing wrong with being hardheaded . . . about the right things.'

"Nobody knew anything about DNA back then," I say. "It hadn't been discovered yet. People knew who they were by the photos they kept; by the Bibles they stuffed with birth certificates, marriage licenses, and the like; by old letters from family and friends; and by stories that made ancestors real individuals with emotions, talents, and values. The stories didn't just tell who begat whom or what person did this or that; they were real-life examples of what is truly important. DNA proof would be nice to have, but it wouldn't define what makes us proud to be who we are. Having James Madison on our family tree is pretty cool, but," I tell my inquisitors, "the pride we feel as a family, as a people, has little to do with a president."

Our stories begin with our names. Mandy's owner gave her a slave name and then listed her among his belongings. That has changed. My daughter, whom my husband and I named Nicole Elise, was brought up knowing that her life and choices are her own. She and her husband, Peter, gave their children names that honor their families. They named their son Peter Lee, after his father and his maternal grandfather, Lee. Their confident, joyful daughter is Madison Lyfe.

When plans for the 2014 reunion of the black Madisons were under way, I asked Nicole whether she was coming.

"Of course," she replied, "and I'm bringing Madison. It's *her* party."

Mandy

I cried when I saw you. You, my baby girl, were streaked with blood, the blood of our African ancestors. You were strong and beautiful, like my mother, my grandmother, my great-grandmother, and the women before her, though your skin was lighter and more golden, glistening like the morning glow on the plains that had nurtured my village for generations. When you arched in my resolute arms, your back and legs felt sturdy, like the trunk and limbs of the tree standing watch over the village where I had learned to fight for the me who I am. Slick with the water from my womb, your hair glowed red as if the promise of the sun were your only equal. When I kissed your smooth little cheek, it tasted salty, like the mighty ocean that had inspired me. You cried, and the drumbeat of your voice foretold the resilience inside you. And I remembered the evening I arrived in this place and the singing I heard in the distance. The melody, I now knew, was a lullaby of possibility for an irrepressible child. I threw back my head and cried hope-filled tears.

Acknowledgments

From the moment I first set foot on Montpelier in 1992, I felt welcomed. Lynne Lewis, director of the archaeology staff at that time, immediately recognized the importance of my research and trusted me to see their most important recent discovery, the remains of the south kitchen. This trust allowed me to walk, literally, in the footsteps of my enslaved ancestor Coreen. During a later visit, Lynne took me to the slave cemetery, where I "found" my beloved Mandy.

The trust, support, and respect I have felt at Montpelier do not end with Lynne. Christian Cotz, director of education and visitor engagement, included me on the advisory board for *A Mere Distinction of Colour,* a permanent exhibit about the importance of Montpelier's slaves to James Madison and, ultimately, to the formation of the nation. I am proud that my image and voice are part of the exhibit. Matthew Reeves, current director of archaeology and landscape restoration, encouraged me to participate in an archaeological dig in search of slave artifacts. I felt like a real archaeologist when the knees of my jeans became coated with Virginia's red soil. In 2017, Matt and Christian bestowed upon me the honor of representing Montpelier on a panel of slave descendants at a symposium held at the University of Virginia. Hannah Scruggs, research associate for the African American Descendants' Project, has taken care to include me in every program for and about Montpelier's slaves and their descendants. When I asked Elizabeth Chew, vice president for museum programs and chief curator, to read a draft of my book, I was concerned she would reject any unfavorable statement about the Madisons, but Elizabeth is a champion of my story, my research, and my writing. Hilarie Hicks, senior research historian, also reviewed my book and offered several important historical details, clarifications, and insights, which augmented the accuracy of the manuscript.

Ann L. Miller, research scientist and historian for the Virginia Transportation Research Council, has been a valuable resource for all sorts of information about the people and history of Orange County and its surrounding areas. Her words brought the people, places, and times of the past into the present for me.

Searching every archive they could get access to, Jimmy Madison and Sean Harley dug up invaluable difficult-to-find bits of information about our family that opened up new avenues of investigation. Thank you, cousins.

To develop my writing skills, I sought out other writers, took classes, and joined critique groups in the Boston area. Special thanks to my first writing instructor, Gail Pool. In her class Writing for Publication at the Radcliffe Seminars — which included fellow students Gloria Jean Gallington and Elizabeth Marcus — I acquired the skills, fortitude, and persistence required to be a writer.

The group that has been part of my life for some twenty years was inspired by our muse, Zora Neale Hurston. We proudly called ourselves "Zora's Girls." Je'Lesia Jones, Renée Gold, Robin Reed,

Carrie Johnson, Ellen Story, and Arlene Weiland are my sisters of the pen. We wrote, critiqued, agonized, laughed, and cried together.

For two years, I was also a member of the critique group, affectionately known as "the Critical Mass," at the Writers' Loft. In an actual loft in Sherborn, Massachusetts, Dave Pasquantonio, Geoff Parker, Jessica Sweet, Tod Dimmick, Ben Williams, Erica Boyce Murphy, Brent Hall, Deborah Mead, and Deborah Norkin, all astute readers and strong writers, offered in-depth and profoundly insightful analyses.

Another group at the Writers' Loft, the Nonfiction Think Tank, which included Nancer Ballard, Elana Varon, and Jui Navare, made me look beyond the trees so that I could see the forest of my book.

My mother had handed me such a wealth of material that I was in danger of overwhelming my readers by attempting to include everything. "Book Doctor" Ellen Szabo helped me to focus the story I wanted to tell, and "Book Architect" Stuart Horwitz guided me toward creating a sound structural foundation upon which to build that multifaceted story.

Paula Young Lee, author of many literary articles and several well-regarded nonfiction books, put in a good word for me with her agents at Inkwell Management, Kimberly Witherspoon and Jessica Mileo. They agreed to read my book and then offered to take me on. Simply stated, Paula, Kim, and Jessica made things happen for me.

Lauren Wein, my pencil-in-hand editor, and her successor Pilar Garcia-Brown asked hard questions about my purpose. The answers helped me to place my personal story within the larger historical framework and to become an active participant in the ongoing conversation.

Tracy Roe, my copyeditor, fact checker, and fellow MD, analyzed my obsession with commas and quotation marks and led me to recovery. My production editor, Jennifer Freilach, calmly handled my episodes of panic and made sure everything was in place and looked just right. The expertise, dedication, and enthusiasm of the publicity, marketing, and sales team, Emma Gordon, Liz Anderson, Hannah Harlow, and Alia Almeida made getting my book out to the public an exciting adventure.

From start to finish, my cheerleading team, Danielle Buckman, Tommy Simms, Carol Hovey, June Tarter, Clemmie Cash, Margot Patricia Forneret, Bonnie Peet, and Ellen Griffiths, kept my spirits high.

Throughout the many years I worked on this book, my husband, Lee, gave me support.

From the moment she was born, my daughter, Nicole, gave me joy.

Indeed, it takes a village.

On June 27, 2009, my mother passed away. I hope she is looking down at me and saying,

I love the way you told our story.

Ruby Laura Madison Wilson (1918–2009), circa 1980

Image Credits

All photos are courtesy of the author, with the exception of: "Henry Green Madison (1840–1912)" and "Henry Green Madison's cabin in Rosewood Park, Austin, Texas," which are used courtesy of Sean Harley; "Jesse Billingsley (1810–1880)," which is used with permission by the San Jacinto Museum of History; and "Portrait of James Madison, Sr. (oil on canvas, 1799)" by Charles Peal Polk used with permission by Belle Grove Plantation, Middletown, Virginia.

Resources

Allgor, Catherine. *Dolley Madison: The Problem of National Unity.* New York: Routledge, 2018.

Appiah, Kwame Anthony, and Henry Louis Gates Jr. *Africana: The Encyclopedia of the African and African American Experience.* New York: Basic Civitas Books, 1999.

Barr, Alwin. *Black Texans: A History of African Americans in Texas, 1528–1995.* Norman: University of Oklahoma Press, 1997.

Bence, Evelyn, ed. *James Madison's Montpelier: Home of the Father of the Constitution.* Orange County, VA: Montpelier Foundation, 2008.

Bergman, Peter. *The Chronological History of the Negro in America.* New York: Harper and Row, 1969.

Blesser, Carol, ed. *In Joy and in Sorrow: Women, Family, and Marriage in the Victorian South, 1830–1900.* New York: Oxford University Press, 1991.

Blumrosen, Alfred W., and Ruth G. Blumrosen. *Slave Nation: How Slavery United the Colonies and Sparked the American Revolution.* Naperville, IL: Sourcebooks, 2005.

Bryant, Linda Allen. *I Cannot Tell a Lie: The True Story of George Washington's African American Descendants.* New York: iUniverse, 2004.

Chambers, Douglas B. *Murder at Montpelier: Igbo Africans in Virginia.* Jackson: University Press of Mississippi, 2005.

Clark, Kenneth M. "James Madison and Slavery." James Madison Museum, 2000.

Cohn, Paul D. *São Tomé.* Bozeman, MT: Burns-Cole, 2005.

Cumbo-Floyd, Andi. *The Slaves Have Names: Ancestors of My Home.* Scotts Valley, CA: CreateSpace, 2013.

Du Bois, W.E.B. *Voices from Within the Veil.* New York: Harcourt, Brace, 1920.

Ellis, Joseph J. *American Sphinx: The Character of Thomas Jefferson.* New York: Random House, 1996.

Galland, China. *Love Cemetery: Unburying the Secret History of Slaves.* New York: Harper One, 2007.

Genovese, Eugene D. *Roll, Jordan, Roll: The World the Slaves Made.* New York: Random House, 1971.

Gordon-Reed, Annette. *The Hemingses of Monticello: An American Family.* New York: W. W. Norton, 2008.

Hale, Thomas A. *Griots and Griottes: Masters of Words and Music.* Bloomington: Indiana University Press, 1998.

Hall, B. C., and C. T. Wood. *The South.* New York: Scribner, 1995.

Hartman, Saidiya. *Lose Your Mother: A Journey Along the Atlantic Slave Route.* New York: Farrar, Straus and Giroux, 2007.

Horton, James Oliver, and Lois E. Horton. *Slavery and Public History: The Tough Stuff of American Memory.* New York: New Press, 2006.

Johnson, Charles, Patricia Smith, and the WGBH Research Team. *Africans in America: America's Journey Through Slavery.* New York: Harcourt, Brace, 1998.

Jordan, Winthrop D. *White Over Black: American Attitudes Toward the Negro.* Chapel Hill: University of North Carolina Press, 1968.

Kelley, Robin D. G., and Earl Lewis, eds. *To Make Our World Anew: A History of African Americans.* Oxford: Oxford University Press, 2000.

Ketcham, Ralph. *James Madison: A Biography.* Charlottesville: University of Virginia Press, 1971.

Madden, T. O., Jr., with Ann L. Miller. *We Were Always Free: The Maddens of Culpeper County, Virginia.* New York: Vintage, 1992.

Miller, Ann. *The Short Life and Strange Death of Ambrose Madison.* Orange County, VA: Orange County Historical Society, 2003.

Miller, John Chester. *The Wolf by the Ears: Thomas Jefferson on Slavery.* Charlottesville: University of Virginia Press, 1991.

Mintz, Sidney W., and Richard Price. *The Birth of African American Culture: An Anthropological Perspective.* Boston: Beacon Press, 1976.

Nelson, Alondra. *The Social Life of DNA: Race, Reparations, and Reconciliation After the Genome.* Boston: Beacon Press, 2016.

Nooter, Mary H. *Secrecy: African Art That Conceals and Reveals.* New York: Museum for African Art, 1993.

O'Reilly, Kenneth. *Nixon's Piano: Presidents and Racial Politics from Washington to Clinton.* New York: Free Press, 1995.

Rainville, Lynn. *Hidden History: African American Cemeteries in Central Virginia.* Charlottesville: University of Virginia Press, 2014.

Smith, James Morton. *The Republic of Letters: The Correspondence Between Thomas Jefferson and James Madison, 1776–1826.* New York: W. W. Norton, 1995.

Stevenson, Brenda E. *Life in Black and White: Family and Community in the Slave South.* New York: Oxford University Press, 1996.

Thomas, Hugh. *The Slave Trade: The Story of the Atlantic Slave Trade, 1440–1870.* New York: Simon and Schuster, 1997.

Thompson, Robert Farris. *Flash of the Spirit: African and Afro-American Art and Philosophy.* New York: Random House, 1983.

Thompson, Robert Farris, and Joseph Cornet. *The Four Moments of the Sun: Kongo Art in Two Worlds.* Washington, DC: National Gallery of Art, 1981.

Van Dantzig, Albert. *Forts and Castles of Ghana.* Accra, Ghana: Sedco Publishing, 1980.

Vlach, John Michael. *Back of the Big House: The Architecture of Plantation Slavery.* Chapel Hill: University of North Carolina Press, 1993.

Wilkins, Roger. *Jefferson's Pillow: The Founding Fathers and the Dilemma of Black Patriotism.* Boston: Beacon Press, 2001.

Williams, Andrea Heather. *Help Me to Find My People: The African American Search for Family Lost in Slavery.* Chapel Hill: University of North Carolina Press, 2012.

Discussion Questions

1. Bettye Kearse introduces us to her family's credo: "Always remember — you're a Madison. You come from African slaves and a president" (12). These words, as Kearse writes, guided her family for nine generations, especially in the antebellum years when her enslaved ancestors used James Madison's name as a tool to help them find family members who had been sold and sent away. Discuss the importance of a name. Can names be used as means of imprisonment or linking or both?

2. Discuss what it means to be a *griot* or *griotte*. How does Kearse react when she learns that she'll be the next *griotte* in her family? What does she foresee as the greatest challenges?

3. What is the significance of the cotillion scene (13–15)? What does it foreshadow?

4. Discuss the role of community in this memoir—both in

Kearse's own life and in the lives of her ancestors. What has community offered them?

5. Kearse writes, "Though many in our family have heard we descend from President Madison and his slaves, only the *griots* know the full account of our ancestors, white and black, in America" (38). Discuss the importance of oral history. What does it provide that written history cannot?

6. Gramps says, "Our white ancestors laid the foundation for this country, but our dark-skinned ancestors built it" (94). Kearse embarks on her own journey in an effort to better understand her ancestry and her family's regard toward it. How might you imagine confronting this reality when some ancestors oppressed and tried to erase others? What are some of the struggles Kearse experiences on her journey?

7. Discuss the significance of the Mandy sections. What insight do Kearse's imaginings of Mandy offer us? What do these sections bring to the book? Why was it important to Kearse to trace Mandy's footsteps?

8. Kearse writes, "If you shake any family tree, a chain will rattle" (143). Discuss what she means by this, especially within the context of chapter 11, 'Visiting.'

9. In chapter 13, "In Search of the President's Son," Kearse describes her archival and scientific research. How has tech-

nology played a role in linking family histories? What are its blind spots?

10. Discuss the realizations that Kearse has after attending a workshop at Montpelier (232–36). What does the inclusion of the words "other persons" in the Constitution signify?

11. Was there a particular person from Kearse's family history who captured your attention and whom you wished you knew more about? What about their story has been lost to time? What would a fuller sense of their story offer to your understanding of history?

12. Why do you think it was important to Kearse to write this book? What message(s) did you take away?

Q&A with Bettye Kearse

Why is oral history important?

Many cultures have age-old traditions of oral history. The stories of the ancestors and the history of a people link past and present, preserving not just a family or a community, but also entire cultures and their values. The stories tell us what kind of people our ancestors were and how we came to be the kind of people we are. American enslavers successfully abolished many African customs, but the tradition of oral history held strong. For many African-American families, like mine, this tradition is all that preserves the legacies our ancestors left for us.

Why was it important to you to visit the places where your ancestors lived?

Mandy was my family's first African ancestor in America, and being her descendant shaped who I am. So, I thought that if I followed Mandy's path from Africa to America, I could better understand this vital part of myself. But I learned that it is impossible to know what it is like to be stolen, to lose everything I know and everyone I love, to stand on an auction block, to be vulnerable to someone else's power, and, finally, to be enslaved for the rest of my life. I can only hope that in my own life I will do justice to Mandy's life and keep memory of her alive.

What was the most surprising discovery you made in your research?

By excluding the words "slave" and "slavery," choosing instead the term "other persons," the framers of the Constitution made concrete the perception that enslaved Africans were lesser. *But*, I was surprised to discover, the Founding Fathers had inadvertently, and counter to their intent, admitted that the nation's enslaved were human, though, perhaps, only three-fifths so.

What do you hope readers take away from your story?

Racism tries to convince African Americans that we are lesser, that we have nothing to be proud of and little to contribute. These myths are far from the truth. America's enslaved people possessed remarkable inner strength and talents. These qualities did not die with them but were passed down to their descendants, enabling us to make remarkable and important contributions to America and the world.